DEBATING SCIENTIFIC KNOWLEDGE AND TECHNOLOGY

EDITED BY

KEITH PARSONS

59 John Glenn Drive
Amherst, New York 14228-2197

Published 2003 by Prometheus Books

Inquiries should be addressed to
Prometheus Books
59 John Glenn Drive
Amherst, New York 14228–2197
VOICE: 716–691–0133, ext. 207
FAX: 716–564–2711
WWW.PROMETHEUSBOOKS.COM

07 06 05 04 03 5 4 3 2 1

Library of Congress Cataloging-in-Publication Data

The science wars : debating scientific knowledge and technology / edited by Keith Parsons.
p. cm. — (Contemporary issues)
Includes bibliographical references.
ISBN 1–57392–994–8 (pbk.)
1. Science—Social aspects. 2. Science and state. I. Parsons, Keith, 1952– II. Contemporary issues (Amherst, N.Y.)

Q175.55 .S295 2003
306.4'5—dc21

2002036989

Printed in the United States on acid-free paper

CONTENTS

INTRODUCTION

FROM GISH TO FISH

Personal Reflections on the Science Wars

KEITH PARSONS

Though their looks may be mild, professors are a combative lot. They battle over issues that others may have a hard time understanding, much less getting upset about. Academics take ideas very seriously—which is good, because someone needs to. Since dialogue and debate are the lifeblood of every field of inquiry, the clash of ideas is usually a sign of intellectual vitality. Sometimes, though, conflicting ideas will involve such fundamental differences in worldview that debate becomes bitter. Throw in some disciplinary chauvinism and a few hyperinflated egos, and a nasty situation can become explosive. Most professorial brawls are tempests in teapots—vehemently contested in learned journals and colloquia, but of little immediate interest to those outside academe. The "science wars" are different; they involve the status of science and technology in our culture.

The scientific and technological achievements of the past century must astonish us: The theory of relativity, quantum mechanics, molecular biology, plate tectonics, digital technology, fiber optics, lasers, genetic engineering . . . the list could be extended over pages. Now, early in a new century and new millennium, we look to the future with both hope and apprehension. My grandfather was a young man when the Wright brothers flew the first airplane; by the time of his death, humans had been to the moon. If science and technology have taken us so far so fast, what will the future be like? The very success of science can make it frightening.

On the other hand, some of science's failures seem as glaring as its successes. Cancer kills hundreds of thousands of Americans annually, despite the billions of dollars and untold time and energy spent since Richard Nixon announced a "war on cancer." We remain dependent on costly and polluting fossil fuels despite bright promises of new, clean, cheap, and endless energy. In general, the future isn't what it used to be. The 1939 World's Fair had the theme "The World of Tomorrow." Its displays predicted that science and technology would give us lives of leisure in sparkling cities with uncongested highways and automated wonders that would banish housework. What happened? Why does our twenty-first-century world look more like *Blade Runner* than The World of Tomorrow? Why, with all of our "time-saving" gizmos, is life so complicated and its pace so insanely accelerated? Why, with life expectancy more than doubled over the last two hundred years, does it seem that we are living longer and enjoying it less?

Clearly, we need a better understanding of science and technology and their place in our society. We need to know just what science can and cannot do for us. We must be able to evaluate new technologies to judge their promise or threat. We need to know how advancing science and technology will affect our health, our way of life, our values, and our religion. Just at this juncture, when our need to understand science and technology is so urgent, we find leading scholars, scientists, and activists embroiled in furious controversies on these very topics. As a way into these issues, I would like to begin with a personal narrative.

I entered the first grade in the fall of 1958. The previous autumn, the Soviet Union had astounded the world by launching *Sputnik,* the first artificial satellite. Triumph after triumph followed for Russian space technology. By contrast, U.S. space efforts were comically inept as rockets fizzled, exploded, and veered off the launchpads in every direction except up. Politicians and pundits wrung their hands and cried that we were losing the "space race." Educators worried that American children were falling behind their Russian counterparts and reacted by greatly enhancing the science and math curriculum. Consequently, I was taught a lot of science early . . . and I loved it.

I was also taught that science and technology had been a force for good in history. Science had been the scourge of obscurantism and superstition, and technology had aided social progress and all that advanced civilization. In my history books, Benjamin Franklin's electrical discoveries figured as prominently as his statesmanship. Thomas Edison, Alexander Graham Bell, Samuel Morse, George Westinghouse, and the Wright brothers were great American heroes. Even more remarkably, though I grew up in the Bible Belt, none of my teachers ever questioned evolution or that the universe was billions of years old. Of course, I had occasionally glimpsed sweaty evangelists thundering against Godless "evil-lution." Once I even came across a hilariously moronic antievolution comic book called "Big Daddy?" with a smug, banana-eating ape on the cover. But the idea that anyone with higher than a third-grade education would think that humans and dinosaurs had inhabited the earth simultaneously was simply inconceivable.

It became conceivable in the early 1980s. In the fall of 1982, while I was a graduate student at Queen's University in Ontario, Canada, Dr. Duane Gish of the Institute for Creation Research came to our campus to present his version of "scientific creationism." Dr. Gish was no Bible-thumping evangelist; he had a Ph.D. in biochemistry and clearly was a very intelligent man. In a lengthy and passionate oration, he denounced evolutionary biology as scientifically worthless and morally pernicious. He was articulate and his arguments were clever, but I recalled the words of the great geneticist Theodosius Dobzhansky, who rightly said that "nothing in biology makes sense except in the light of evolution." Could biological science be a clay-footed giant? Worse, Dr. Gish was a "young earth" creationist who held that the universe is no more than 10,000 years old. But the central theories and best-confirmed data of astrophysics and cosmology imply that the universe is billions of years old. Some years later I learned that Dr. Gish had written a book claiming that dinosaurs had been in the Garden of Eden (where *T. Rex* had been a harmless herbivore before the Fall of Adam and Eve) and were on the ark with Noah. Clearly, if Dr. Gish was right, very much of science—not just evolution—must be wrong. So much would have to be wrong that we ought to regard the entire course of science since the sixteenth century as something of a very bad joke.

Creationism flourished in the conservative atmosphere of the Reagan years. Like Isaac Asimov, one of the heroes of my youth, I was appalled that in the 1980s scientists had to fight once again battles that they thought they had won before the 1880s. How, near the end of a century of so much scientific progress, could some of the best-confirmed results of science (e.g., that evolution had occurred and that the earth is billions of years old) still provoke angry rejection? Still, creationism was an element of the Religious Right movement, which also gave us the Moral Majority and the Christian Coalition. This movement occasionally gained political clout, but for nearly all academics, its intellectual status was zero. In fact, the biggest boost for creationism was the silence of many mainstream scientists who did not take it seriously enough to bother refuting it. Despite all their "scientific" jargon and flaunting of credentials, antievolutionists were often still dismissed as ignorant hayseeds, an image that had stuck to them since the 1925 Scopes trial.

If practicing scientists could ignore the creationists, which they largely did, how should they react when holders of prestigious academic appointments began saying things about science that sounded as extreme as any of Dr. Gish's claims? How, for instance, should they react when a leading feminist theorist referred to Newton's *Principia Mathematica*—perhaps the most important scientific work ever produced—as a "rape manual"? Some feminists even claimed that the scientific ideals of objectivity and impartiality were manifestations of patriarchy. I recall my own astonishment at a philosophical colloquium when a young feminist opined that "objectivity is what a man calls his subjectivity" (she did not say whether her own claim was being asserted as an objective truth).

Then I found out that the feminist critics were only the vanguard and that many others were involved in the radical critique of the traditional authority and status of science. For instance: (1) Sociologists of knowledge propose that science is a "social construct," that is, scientists do not make discoveries; their "facts" are artifacts of the social and linguistic conventions of scientific communities. "Nature" and "reality" are therefore simply whatever scientists agree to regard as natural or real. Further, the methods of science are not objective means of confirming hypotheses about an external world but arbitrary "rules of the game," which serve a merely rhetorical purpose. (2) "Postmodernist" literary critics issue nebulous critiques of science. It is hard to say just what their points are since they protect themselves with the formidably obscure verbal armor of "in" theorists such as Jean-François Lyotard, Jean Baudrillard, Michel Foucault, and Jacques Derrida. However opaque their rhetoric, their aim is clear—to cut science down to size, to debunk scientific authority, and to display scientific reasoning as just another form of discourse, no more "objective" or "rational" than any other. (3) New and ferocious Luddites challenge the latest technologies. Activists litigate against genetic engineering in the courts and are campaigning against biotechnology in the media. In short, the aims, methods, applications, and even the ideals of science have been radically challenged.

At first, scientists ignored the critiques of their colleagues in the humanities and the social sciences. Scientists are very busy, and generally their attitude toward lay critics, as one scientist pointedly expressed it, is to "let the asses bray." In 1994 the dam burst. In that year a classic of polemical literature was published: *Higher Superstition: The Academic Left and Its Quarrels with Science,* written by biologist Paul R. Gross and mathematician Norman Levitt. This unabashedly pugnacious work pulled no punches in taking on the academic science critics. Gross and Levitt gleefully dwelt on the rhetorical extravagances of "postmodernist" literary critics and the scientific ignorance of Marxist and feminist theorists. Social constructivist sociologists, radical environmentalists, and extremist anti-AIDS activists also took it on the chin. Naturally, those criticized on the "academic left" fired back, and so the science wars were joined.

One of the readers of Gross and Levitt's work was a physicist named Alan Sokal. He decided that debunking the social constructivists was not enough and that more drastic action was required. He composed "Transgressing the Boundaries," an essay intentionally loaded with trendy gibberish, unsupported assertions, and scientific howlers. He sent this parody—as an apparently serious submission—to *Social Text,* a leading periodical of the "academic left." In fact, the founder of *Social Text* was Stanley Aronowitz, a Marxist sociologist who regards science as the embodiment of capitalist power relations. *Social Text* published the essay as a serious article, and Sokal immediately revealed the hoax. He concluded that his successful sting had revealed the deep ignorance of the science critics and their general intellectual laziness and dishonesty. Embarrassed and outraged, the editors of *Social Text* and other left-wing academics fired broadsides back at Sokal, and so the science wars reached a crescendo.

One of the most potent rebukes to Sokal was written by Stanley Fish, a leading literary scholar and, at the time of the hoax, head of the Duke University Press (publisher of *Social Text*). In an essay published in the *New York Times* titled "Professor Sokal's Bad Joke," Fish took Sokal to task. He charged that Sokal had undermined the trust, essential for any field of intellectual endeavor, that all inquirers are engaged in good faith inquiry and not constructing spiteful Trojan horses to humiliate colleagues. According to Fish, such skullduggery calls into question the integrity of the whole academic process and distracts us from serious issues. Fish further endorsed and defended the social constructivist view that Sokal had parodied.

So the critics of science now range from Gish to Fish. From creationists on the right to postmodernists on the left, science has been subjected to intense critical scrutiny. The purpose of the present anthology is to present some of the arguments of the science critics and the rebuttals these have provoked. I have striven to present the most cogent and articulate statements of the science critics and of those who replied to them. My aim is not to present diatribes, which will only add fuel to the conflagration, but to focus on the serious underlying issues. My hope is that readers will come away with a determination to think through for themselves the nature of science and its place in our society.

My own views of the science wars are expressed in my book *Drawing Out Leviathan: Dinosaurs and the Science Wars* (Indiana University Press, 2001). There I argue that the constructivist and postmodernist critiques of science do not amount to much. In my view, their points are hackneyed or trivial when true, and false or incoherent when they promise something interesting. I say this not to bias the reader but, on the contrary, because I think readers have every right to know where the editor stands. I shall strive to be fair, but a pose of total neutrality would be disingenuous. So informed, readers can take care to make up their minds for themselves.

One thing seems certain: Science is here to stay. We can no more go back to a prescientific, mythological worldview than we can shed our skins.

So, we have to figure out what to do about science. My hope is that the present volume will introduce readers to the issues and encourage them to put their minds to this task. There are few issues more important for the future of our civilization than the place of science in our culture. Albert Einstein said that the known is infinitesimal when compared to the unknown, but that what we do know is our most precious possession. Is it? Does science produce objective, impartial knowledge or, like fashion and politics, do social and cultural influences determine its course? Hard thinking will be required to answer these questions. As editor, I shall be handsomely repaid if this work motivates some to this task.

PART 1

THE CONSTRUCTIVIST CHALLENGE

INTRODUCTION

In the early decades of the twentieth century, a young scientist named Alfred Wegener proposed a startling hypothesis. He suggested that the continents had moved—that over the vast stretches of geological time, the continents had shifted great distances across the globe. His evidence was intriguing. He pointed out the coastlines of different continents that seem to fit together like pieces of a jigsaw puzzle (children looking at a world map quickly notice how South America "fits" into Africa). He noted that there were many geological features, such as mountain ranges composed of distinctive types of rock, that correspond to the same sort of ranges on other continents, and would form one continuous range if the continents were moved together. Wegener showed that some fossils were found only in pockets extending inland from the coasts of one continent and in similar pockets on another continent where the two landmasses were now separated by a broad ocean. However, if the coastlines were imagined joined together, one continuous habitat for the fossil creatures would be formed. These and other geological oddities suggested that the continents had drifted to their present positions from earlier, very different configurations.

Hardly any respected geologist took Wegener's proposal seriously. First, he was a meteorologist, not a geologist, and so had no professional standing in the field. Worse, he gave no account of how continents, composed mostly of lighter rock, could move through the dense bedrock of ocean floors. Wegener's theory was derided as "geopoetry" and held up to contempt when orthodox geologists bothered to notice it. Then, in the 1960s, a remarkable thing happened. Within that decade continental drift went from heresy to orthodoxy within the geological community. Within a few years, it was not the advocates but rather the opponents of drift hypotheses who were regarded as the oddballs. Continental drift hypotheses were incorporated into the theory of plate tectonics, which became, and remains, the central theory of geology. Wegener (who died in 1930) was no longer remembered as a pariah, but as an amazingly prescient theorist.

What happened? How could a hypothesis (almost) universally rejected by geologists become (almost) universally accepted in just a few years? Such amazing reversals are well known in the history of science. Contrary to legend, most of the opposition to Charles Darwin's *Origin of Species* (1859) issued from scientists, not religious leaders (many of whom found evolution acceptable or even congenial). Yet before Darwin's death in 1882, the fact of evolution (if not Darwin's explanation of the process) had become the consensus view among scientists. Similarly, though Galileo faced scientific as well as ecclesiastical opposition, heliocentrism was the accepted view by 1650. Sometimes a new theory takes over a whole field so that, as philosopher and historian of science Thomas Kuhn put it, a "paradigm shift" occurs, that is, the field redefines itself in terms of that theory. What happens when a theory, maybe a new one or maybe an old one that had been kicked around for years, takes a whole field of science by storm?

Scientists and philosophers of science offer a standard account of how controversial hypotheses come to be accepted: Science is a preeminently rational enterprise that judges claims on their merits. When a hypothesis has been around for a long time and suddenly becomes acceptable, this is because (a) a new theory has been proposed and strongly confirmed that implies the hypothesis, and/or (b) massive new and cogent evidence suddenly accrues in favor of the hypothesis. Thus, continental drift (a) was incorporated into the broader theory of plate tectonics, which explained how continents could float on denser crustal plates that shift and rotate over geological time. Also, (b) it was supported by much new evidence, including compelling indications that seafloors spread—pushing the continents apart—as new material wells up from midocean ridges.[1] In short, hypotheses are promoted (or demoted) on the basis of their rational credentials: How well do they stack up to the evidence? Can they be incorporated into well-confirmed background theories?

Of course, scientists and philosophers have always known that the story is more complicated than this. Scientists, even the greatest of them, are not unfeeling calculating machines, but subject to all the passions and foibles of other humans. Consider Isaac Newton, perhaps the most astonishing scientific genius of all time. Newton was a man of towering—and touchy—ego, who was not above engaging in acrimonious quarrels over priority (he, or rather his proxies, and philosopher Gottfried Leibniz engaged in a famous spat over who had first formulated the calculus). Worse, fellow scientists who incurred his wrath found that Newton would use his enormous reputation and influence to suppress their work. Impatient, irascible, and eccentric, Newton shaped the course of science with his personality as well as his transcendent genius.

Scientists can be eccentric as individuals, but when they come together to form communities (and science is done by *communities;*

the lone wolf is practically unknown), those communities, like all other human groups, are rife with politics. For instance, the proverbially greater importance of "who you know" over "what you know" applies in science as well. The best way for the fledgling scientist to promote his or her own theory is to form an alliance with a big shot in the field. Different allied groups often come into conflict, and internecine conflicts can be quite nasty. When scientists clash, they do not politely submit data for dispassionate analysis. Overheated rhetoric fills the air; reputations are tarnished and careers are sometimes ruined. A paper published by physicist Luis Alvarez and coauthors in 1980, one proposing a radical new theory of mass extinction, precipitated a series of furious exchanges. Soon all dignity was lost as scientific debate descended into name-calling. Further, scientists have vested interests like everyone else. A scientist who enjoys a cushy job in a polluting industry is often less receptive to evidence of the bad effects of those pollutants.

Clearly, then, the course of science is often influenced by "unscientific" factors. Those who practice science have known this all along. But how far does the influence of such nonrational factors extend? Traditionally it has been held that scientists employ methods that permit science, at least in the long run, to transcend politics and vested interest and give *nature* the final say in which theories we accept. Thus, experiments are rigorously designed to exclude all possible sources of bias, and procedures, such as double blinding, attempt to eliminate even the most subtle errors. When a new hypothesis is evaluated, strict statistical tests must be passed before the null hypothesis is abandoned. Measurements of great precision are employed, and skeptical scientists must be satisfied that objective, reliable procedures have been used to generate results. In general, great care is taken to make sure that no hypothesis is long entertained if it is inconsistent with natural fact. As the old saying goes, science proposes and nature disposes.

Recently, some scholars have radically challenged the traditional view of how hypotheses are accepted in science. In the 1970s, sociologist Bruno Latour decided to study scientists in their native culture just as an anthropologist would do with a rain forest tribe.

He took a menial job at the Salk Institute, the laboratory of the famous immunologist Jonas Salk and a major center for biomedical research. He observed scientists as they moved through their daily routine, listened in on their conversations, and cataloged their behaviors. In 1979 he coauthored a book with Steve Woolgar titled *Laboratory Life: The Construction of Scientific Facts.* As the subtitle of their book indicates, Latour and Woolgar took a very nontraditional view of how "facts" are established in science. In their view, scientific "facts" are not discovered but constructed. Scientific facts are really artifacts, artificial products of a social process, not the revelation of an external, independent, natural reality. Indeed, the terms "nature" and "reality" can only indicate what scientists agree to regard as natural and real, and do not refer to states of affairs that exist prior to inquiry.

But what about the methods of science that are supposed to guarantee that scientific representations are rigorously constrained by *nature,* and so not merely figments of the scientific imagination? For Latour and Woolgar, scientific methods are not reliable means of learning about external reality; they merely provide *rhetorical* ammunition that scientists employ to persuade other scientists. Thus, the strict protocols guiding experiments, and the enormous care taken in measurements and in accurate recording of information, are like the elaborate rituals of the primitive shaman. The shaman performs his rituals carefully so that the gods will hear him. Likewise, a researcher who wants to achieve credibility in the eyes of other scientists must perform these rites well or they will not listen. If, however, the other scientists are shown that the prescribed rituals have been followed precisely, they might be persuaded to elevate the controversial claim to "fact." In short, the copious references to techniques and methods in scientific papers are just *rhetorical devices* employed in the process of negotiation and persuasion whereby scientists get colleagues to accept their views.

For Latour and Woolgar this whole business is dubious if not downright sleazy. When what they call an "inversion" occurs—when a scientific community agrees to stop regarding a claim as mere hypothesis and elevates it to the status of fact—the process whereby consensus was reached is obscured. Scientists conveniently forget that the fact is an artificial product, achieved by rhetoric and negotiation, and speak of it as though it had simply been there, in external reality, all along. Thus does scientific fantasy become enshrined as fact before which all are expected to bow. Biologist Matt Cartmill uses provocative language but gives an essentially accurate statement of Latour and Woolgar's constructivist view of science:

> The philosophy of social constructivism claims that the "nature" that scientists pretend to study is a fiction cooked up by the scientists themselves—that, as Bruno Latour puts it, natural objects are the *consequence* of scientific work rather than its *cause.* In this view, the ultimate purpose of scientists' theories and experiments is not to understand or control an imagined "nature," but to provide objective-sounding justifications for exerting power over other people. As social constructivists see it, science is an imposing but hollow Trojan horse that conceals some rather nasty storm troopers in its belly.[2]

In later works published in the 1980s, such as *Science in Action* and *The Pasteurization of France,* Latour extended his critique. In these later works, even the rhetorical role of scientific methods was deemphasized and the importance of scientific "networks" was highlighted. For Latour, scientists trying to convince skeptical and even antagonistic colleagues are engaged in a form of ritualized warfare. In a fight it is useful to have as many powerful allies as possible. Scientists therefore seek to form a network of powerful alliances with as many prominent and capable colleagues as they can. If a scientist can command sufficient resources in the form of money, institutional support, and the backing of influential friends, then opponents can simply be outgunned. Would-be dissidents cannot match the forces arrayed against them, and so must retire from the field. Latour does not deny that there are "rational" means for assessing hypotheses. However, in the actual rough-and-tumble process of science, invoking those rules would be comically inept, like preaching the Sermon on the Mount in a barroom brawl.

Latour and Woolgar's book was one of the founding documents of the burgeoning field of Science and Technology Studies (STS). This field follows Latour and Woolgar's lead in applying a social constructivist methodology to the study of science. Another of the most important works in STS was Steven Shapin and Simon Schaffer's *Leviathan and the Air-Pump* (1985). This work examines the work of seventeenth-century scientist Robert Boyle, whose experiments with the air pump helped establish the experimental method in science. They trace the conflict between Boyle and philosopher Thomas Hobbes (infamous author of *Leviathan*) who vehemently rejected the experimental approach. Shapin and Schaffer minutely examine the social, political, and religious context of the Boyle/Hobbes controversy. They conclude that Boyle did not win because he had the stronger case, but simply because he curried favor with the power brokers and became an establishment insider. Hobbes, on the other hand, was regarded as an infidel and an otherwise dubious character and was excluded from the sources of patronage and support. Thus Boyle came to be recognized as the scientific authority and Hobbes was dismissed as a crackpot.

Shapin and Schaffer draw a general moral from this case. They conclude that scientific methods are not adopted because they are recognized as better ways of doing science but because they serve local political or social agendas. The upshot is that scientific methods are arbitrary "rules of the game" adopted by scientific communities because of historical contingencies and not because they are more reliable or rigorous. If the very methods, standards, and values of science are thus arbitrary and contingent, then what counts as "rationality," "objectivity," or "evidence" must be socially constructed through and through. What seems to follow is an extreme skepticism about any and all scientific claims. Physical reality, whatever that might be, has a negligible bearing on what we believe; we ourselves are responsible for what we think we know.

Now, the scientifically inclined reader may be tempted to dismiss social constructivism as nonsense, even lunacy. Surely no sane person can deny that there is an external world or, despite the skeptical puzzles that have entertained generations of philosophers, that our senses put us in touch with some of its features. Some defenders of social constructivists reply: Of course no sane person would deny these things, and the constructivists may sometimes be flippant or even outrageous, but they are not crazy. So, constructivism should not be interpreted as a form of radical skepticism that severs our theories from natural reality. Rather, say the defenders, constructivism is a methodology that systematically directs attention to the *human* aspects of science.

But the constructivists really do say things that seem to entail extreme skepticism about science or even about an external, non-social world. If they do not mean these things, they should not say them. I shall leave it to the reader to decide just how seriously to take constructivist claims. They seem very serious to me, and they have become so influential that they cannot simply be ignored. In fact, the dominant paradigm in the sociology of science is now constructivist, and constructivism is highly influential among historians of science as well.

The readings begin with selections from Latour and Woolgar's *Laboratory Life* and Shapin and Schaffer's *Leviathan and the Air-Pump.* Because these selections might be somewhat difficult for those becoming acquainted with them, the next reading is an extended, introductory-level exposition of these works by Robert Klee from his book *Introduction to the Philosophy of Science: Cutting Nature at Its Seams.* Klee also evaluates these works and offers a number of thoughtful criticisms. The section concludes with a selection from Paul R. Gross and Norman Levitt's *Higher Superstition,* their 1994 counterblast to the science critique of the "academic left." Gross and Levitt argue that Shapin and Schaffer fail to acknowledge that Hobbes was in fact a mathematical crackpot whose pretensions had been deflated by some of Boyle's allies in the Royal Society. So, Hobbes was his own worst enemy in his efforts to get his critique taken seriously. In other words, the rejection of Hobbes's ideas was not merely a function of his unpopular politics or his reputation as an infidel, but had a rational basis.

NOTES

1. See Naomi Oreskes, ed., *Plate Tectonics: An Insider's Guide to the Modern Thoery of the Earth* (Boulder, Colo.: Westview, 2001).

2. Matt Cartmill, review of *Mystery of Mysteries: Is Evolution a Social Contract?* by Michael Ruse, *Reports of the National Center for Science Education* 19, no. 5 (1999): 49–50.

ONE
FACTS AND ARTIFACTS

BRUNO LATOUR
AND STEVE WOOLGAR

The paradox associated with the term *fact* can have two contradictory meanings. On the one hand, our quasi-anthropological perspective stresses its etymological significance: a fact is derived from the root *facere, factum* (to make or to do). On the other hand, fact is taken to refer to some objectively independent entity which, by reason of its "out there-ness" cannot be modified at will and is not susceptible to change under any circumstances. The tension between the existence of knowledge as pregiven and its creation by actors has long been a theme which has preoccupied philosophers[1] and sociologists of knowledge. Some sociologists have attempted a synthesis of the two perspectives,[2] but usually with somewhat unsatisfactory results. More recently, sociologists of science have convincingly argued the case for the social fabrication of science.[3] But despite these arguments, facts refuse to become sociologized. They seem able to return to their state of being "out there" and thus to pass beyond the grasp of sociological analysis. In a similar way, our demonstration of the microprocessing of facts is likely to be a source of only temporary persuasion that facts are constructed. Readers, especially practicing scientists, are unlikely to adopt this perspective for very long before returning to the notion that facts exist, and that it is their existence that requires skillful revelation.[4] In the last part of this chapter, therefore, we discuss the source of this resistance to sociological explanation. It is little use arguing the feasibility of the strong program of the sociology of knowledge if we cannot understand why it seems systematically absurd to make such an argument. As Kant[5] advised, it is not enough merely to show that something is an illusion. We also need to understand why the illusion is necessary.

In the case of TRF [thyrotropin releasing factor], we showed when and where the metamorphosis between statement and fact took place. By the end of 1969, when Guillemin and Schally formulated the statement that TRF is Pyro-Glu-His-Pro-NH_2, no one was able to raise any further objections to this claim. Laboratories with no interest in the nine-year saga of the emergence of TRF proceeded from this statement merely by citing papers published at the end of 1969. For them the statement was sufficient basis on which to place an order for the synthetic material which promised to decrease the noise of the assays in which they were engaged. From the point of view of the borrowers, the traces of production of the established fact were uninteresting and irrelevant. Five years later, even the names of the "discoverers" of TRF were of no consequence. . . .

We have been careful to point out that our determination of the point of stabilization, when a statement rids itself of all determinants of place and time and of all reference to its producers and the production process, did not depend on our assumption that the "real TRF" was merely waiting to be discovered and that it finally became visible in 1969. TRF might yet turn out to be an artifact. For example, no arguments have yet been advanced which are accepted as proof that TRF is present in the body as Pyro-Glu-His-Pro in "physiologically significant" amounts. Although it is accepted that synthetic Pyro-Glu-His-Pro is active in assays, it has not yet been possible to measure it in the body. The negative findings of attempts to establish the logical significance of TRF have thus far been attributed to the insensitivity of the assays being used rather than to the possibility that TRF is an artifact. But some further slight change in context may yet favor the selection of an alternative interpretation and the realization of this latter possibility. The point at which stabilization occurs depends on prevailing conditions within a particular context. It is characteristic of the process of fact construction that stabilization entails the escape of a statement from all reference to the process of construction.

Facts and artifacts do not correspond respectively to true and false statements. Rather, statements lie along a continuum according to the extent to which they refer to the conditions of their construction. Up to a certain point on this continuum, the inclusion of reference to the conditions of construction is necessary for purposes of persuasion. Beyond this point, the conditions of construction are either irrelevant or their inclusion can be seen as an attempt to undermine the established factlike status of the statement. Our argument is not that facts are not real, nor that they are merely artificial. *Our argument is not just that facts are socially constructed. We also wish to show that the process of construction involves the use of certain devices whereby all traces of production are made extremely difficult to detect.* Let us look more closely at what takes place at the point of stabilization.

From their initial inception, members of the laboratory are unable to determine whether statements are true or false, objective or subjective, highly likely or quite probable. While the agonistic process is raging, modalities are constantly added, dropped, inverted, or modified. Once the statement begins to stabilize, however, an important change takes place. *The statement becomes a split entity.* On the one hand, it is a set of words which represents a statement about an object. On the other hand, it corresponds to an object in itself which takes on a life of its own. It is as if the original statement had projected a virtual image of itself which exists outside the statement.[6] Previously, scientists were dealing with statements. At the point of stabilization, however, there appears to be both objects *and* statements about these objects. Before long, more and more reality is attributed to the object and less and less to the statement *about* the object. Consequently, an inversion takes place: the object becomes the reason why the statement was formulated in the first place. At the onset of stabilization, the object was the virtual image of the statement; subsequently, the statement becomes the mirror image of the reality "out there." Thus, the justification for the statement TRF is Pyro-Glu-His-Pro-NH_2 is simply that "TRF *really is* Pyro-Glu-His-Pro-NH_2." At the same time, the past becomes inverted. TRF has been there all along, just waiting to be revealed for all to see. The history of its construction is also transformed from this new vantage point: the process of construction is turned into the pursuit of a single path which led inevitably to the "actual" structure. Only through the skills and efforts of "great" scientists could the setbacks of red herrings and blind alleys be overcome and the real structure be revealed for what it was.

Once splitting and inversion have occurred, even the most cynical observers and committed relativists will have difficulty in resisting the impression that the "real" TRF has been found, and that the statement mirrors reality. The further temptation for the observer, once faced with one set of statements and one reality to which these statements correspond, is to marvel at the perfect match between the scientist's statement and the external reality.[7] Since wonder is the mother of philosophy, it is even possible that the observer will begin to invent all kinds of fantastic systems to account for this miraculous *adequatio rei et intellectus.* To counter this possibility, we offer our observations of the way this kind of illusion is constructed within the laboratory. It is small wonder that the statements appear to match external entities so exactly: they are the same thing.

Our contention is that the strength of correspondence between objects and statements about these objects *stems from the splitting and inversion of a statement within the laboratory context.* This contention can be supported in three ways. Firstly, there are severe difficulties in adequately describing the nature of the "out there-ness" in which objects are said to reside because descriptions of scientific reality frequently comprise a reformulation or restatement of the statement which purports to "be about" this reality. For example, it is said that TRF is Pyro-Glu-His-Pro-NH_2. But the further description of the nature of the TRF "out there" hinges on the repetition of this statement and so involves tautology. Lest the reader thinks this is an unwarranted caricature of the realist position, it is worth quoting from an argument for a "realist theory of science." In essence, the position advocated here is that no theory of science is possible without what are referred to as "intransitive objects of scientific knowledge."

> We can easily imagine a world similar to ours, containing the same intransitive objects of scientific knowledge, but without any science to produce knowledge of them in such a world, which has occurred and may come again, reality would be unspoken for and yet things would not cease to act and interact in all kinds of ways. In such a world the tides would still turn and metals conduct electricity in the way that they do, without a Newton or a Drude to produce our knowledge of them. The Widemann-Franz law would continue to hold although there would be no-one to formulate, experimentally establish or deduce it. Two atoms of hydrogen would continue to combine with one atom of oxygen and in favorable circumstances osmosis would continue to occur.[8]

The author adds that these intransitive objects are "quite independent of us."[9] He then continues with a striking confession: "They are not unknowable, because, as a matter of fact, quite a bit is known about them."[10] Quite a bit indeed! The marvel of the author for the independence of reality belies its initial construction. Moreover, the ontological status accorded these independent objects is enhanced by the vague terms in which they are described. For example, the statement that "metals conduct electricity in the way they do" implies a complexity beyond the scope of present discussion and, by implication, available only to concerted efforts toward the pursuit and revelation of the reality which gives rise to the description provided here.[11] The author can only recall the reality of the Widemann-Franz laws through the use of eponymy. In addition, he wisely confines his discussion to physics, and to pre-Newtonian physics at that. Perhaps the "independence" of "intransitive objects of scientific knowledge" would seem less unproblematic in relation to more recently constructed phenomena, such as chromosomes or non-Newtonian physics. The realist position, exemplified by the above, centers on a tautological belief whereby the nature of independent objects can only be described in the terms which constitute them. Our preference is for the observation of the processes of splitting and inversion of statements which make these kinds of beliefs possible.

Scientists themselves constantly raise questions as to whether a particular statement "actually" relates to something "out there," or whether it is a mere figment of the imagination, or an artifact of the procedures employed. It is therefore unrealistic to portray scientists busily occupying themselves with scientific activity while leaving debates between realism and relativism to the philosophers. Depending on the argument, the laboratory, the time of year, and the currency of controversy, investigators will variously take the stand of realist, relativist, idealist, transcendental relativist, skeptic, and so on. In other words, the debate about the paradox of the fact is not the exclusive privilege of the sociologist or philosopher. It follows that attempts to resolve the essential differences between these positions is merely to engage in the same kind of debates as the subjects of study, rather than to understand how debates get resolved and positions taken as practical and temporary achievements. As Marx[12] put it:

> [T]he question of knowing if human thought is able to reach an objective truth is not a theoretical but a practical question. It is by practice that man ought to prove the truth, that is, the reality and the power of the something *beyond* his thought.

An important task for the sociologist is to show that the construction of reality should not be itself reified. This can be shown by considering all stages of the process of reality construction and by resisting the temptation to provide a general explanation for the phenomenon.

Perhaps the most forceful argument for the occurrence of splitting and inversion is the existence of artifacts. A modification in the local context of the laboratory may result in the use of a modality whereby an accepted statement becomes qualified or doubted. This yields perhaps the most fascinating observation to be made in the laboratory—the *deconstruction* of reality. The reality "out there" once again melts back into a statement, the conditions of production of which are again made explicit. We have already given a number of examples of this deconstruction process. The existence of a moiety for TRF was taken as fact for a few years and was almost regarded as reality before it faded away and was found to be an artifact of the purification process. Sometimes the status of statements changed from day to day, even from one hour to the next. The factual status of one substance, for instance, varied dramatically over a period of a few days.[13] On Tuesday, a peak was thought to be the sign of a real substance. But on Wednesday the peak was regarded as resulting from an unreliable physiograph. On Thursday, the use of another pool of extracts gave rise to another peak which was taken to be "the same." At this point, the existence of a new *object* was slowly solidifying, only to be dissolved again the following day. At the frontier of science, statements are constantly manifesting a double potential: they are either accounted for in terms of local causes (subjectivity or artifact) or are referred to as a thing "out there" (objectivity and fact).

While one set of agonistic forces pushes statement toward fact-like status, another set pushes it toward artifact-like status. . . . The local status of a statement at any time depends on the resultant of these forces. The construction and dismantling of the same statement can be monitored by direct observation, so that what was a "thing out there" can be seen to fold back into a statement which is referred to as a "mere string of words," a "fiction," or an "artifact."[14] The importance of observing the transformation of a statement between fact-like and artifact-like status is obvious: if the "truth effect" of science can be shown both to fold and unfold, it becomes much more difficult to argue that the difference between a fact and an artifact is that the former is based on reality while the latter merely arises from local circumstances and psychological conditions. The distinction between reality and local circumstances *exists only after* the statement has stabilized as a fact.

To summarize the argument in another way, "reality" cannot be used to explain why a statement becomes a fact, since it is only after it has become a fact that the effect of reality is obtained. This is the case whether the reality effect is cast in terms of "objectivity" or "out-there-ness." It is *because* the controversy settles, that a statement splits into an entity and a statement about an entity; such a split never precedes the resolution of controversy. Of course, this will appear trivial to a scientist working on a controversial statement. After all, he does not wait in hope that TRF will pop up at a meeting and finally settle the controversy as to which amino acids it comprises. In this work, therefore, we use the argument as a methodological precaution. Like scientists themselves we do not use the notion of reality to account for the stabilization of a statement, because this reality is formed as a consequence of this stabilization.[15]

We do not wish to say that facts do not exist nor that there is no such thing as reality. In this simple sense our position is not relativist. Our point is that "out-there-ness" is the *consequence* of scientific work rather than its *cause.* We therefore wish to stress the importance of *timing.* By considering TRF in January 1968, it would be easy to show that TRF is a contingent social construction, and moreover, that scientists themselves are relativists in that they are very aware of the possibility of their constructing a reality which could be an artifact. On the other hand, analysis in January 1970 would reveal TRF as an object of nature which had been discovered by scientists, who, in the meantime, had metamorphosed into hardened realists. Once this controversy has settled, reality is taken to be the cause of this settlement; but while controversy is still raging, reality is the consequence of debate, following each twist and turn in the controversy as if it were the shadow of scientific endeavor.[16]

It could be objected that there are other grounds for accepting the reality of a fact apart from the cessation of controversy. For example, it could be argued that the efficacy of a scientific statement outside the laboratory is sufficient basis for accepting its correspondence with reality.[17] A fact is a fact, one could say, because it works when you apply it outside science. This objection can be answered in the same way as the objection about the equivalence of a statement with the thing out there: observation of laboratory activity shows that the "outside" character of a fact is itself the *consequence* of the laboratory work. In no instance did we observe the independent verification of a statement produced in the laboratory. Instead, we observed the *extension* of some laboratory practices to other arenas of social reality, such as hospitals and industry.

This observation would be of little weight if the laboratory was concerned exclusively with so-called basic science. However, our laboratory had many connections with clinicians and with industry through patents.[18] Let us consider one particular statement: "Somatostatin blocks the release of growth hormones as measured by radioimmunoassay." If we ask whether this statement works outside science, the answer is that the statement holds in every place where the radioimmunoassay has been reliably[19] set up. This does not imply that the statement holds true *everywhere,* even where the radioimmunoassay has *not* been set up. If one takes a blood sample of a hospital patient in order to determine whether or not somatostatin lowers the level of the patient's growth hormone, there is no way of answering this question without access to a radioimmunoassay for somatostatin. One can *believe* that somatostatin has this effect and even claim by induction that the statement holds true absolutely, but this amounts to belief and a claim, rather than to a proof.[20] Proof of the statement necessitates the extension of the network in which the radioimmunoassay is valid, to make part of the hospital ward into a laboratory annex in order to set up the same assay. It is impossible to prove that a given statement is verified outside the laboratory since the very existence of the statement depends on the context of the laboratory. We are not arguing that somatostatin does not exist, nor that it does not work, but that it cannot jump out of the very network of social practice which makes possible its existence.

There is nothing especially mysterious about the paradoxical nature of facts. Facts are constructed in such a way that, once the controversy settles, they are taken for granted. The origin of the paradox is in the lack of observation of scientific practices; when an observer considers that the structure of TRF is Pyro-Glu-His-Pro-NH_2 and then realizes that the "real" TRF is also Pyro-Glu-His-Pro-NH_2, he marvels at this magnificent example of correspondence between man's mind and nature. But closer inspection of the processes of production reveals this correspondence to be much more earthy and less mysterious: the thing and the statement correspond for the simple reason that they come from the same source. Their separation is only *the final stage in the process of their construction*. Similarly, many scientists and nonscientists alike, marvel at the efficacy of a scientific fact outside science. How extraordinary that a peptidic structure discovered in California works in the smallest hospital in Saudi Arabia! For one thing, it only works in well-equipped clinical laboratories. Considering that the same set of operations produces the same answers, there is little to marvel at:[21] if you carry out the same assay you will produce the same object.[22]

By this introduction to the microprocesses of the fact production, we have tried to show that a close inspection of laboratory life provides a useful means of tackling problems usually taken up by epistemologists; that the analysis of these microprocesses does not in any way require the a priori acceptance of any special character of scientific activity; and finally that it is important to eschew arguments about the external reality and outside efficacy of scientific products to account for the stabilization of facts, because such reality and efficacy are the consequence rather than the cause of scientific activity.

NOTES

1. Gaston Bachelard, *Le Matérialisme Rationnel* (Paris: PUF, 1953).

2. Peter L. Berger and Thomas Luckmann, *The Social Construction of Reality* (Harmondswoth: Penguin, 1971).

3. For example, David Bloor, *Knowledge and Social Imagery* (London: Routledge and Kegan Paul, 1976); Harry M. Collins, "The Seven Sexes: A Study in the Sociology of a Phenomenon or the Replication of Experiments in Physics," *Sociology* 9 (1975): 205–24; Karin Knorr, "Producing and Reproducing Knowledge: Destructuve or Constructive," *Social Science Information* 16 (1978): 669–96.

4. Of course, the adoption of this perspective was a practical necessity. The participants were themselves very much aware that they were engaged in construction.

5. Immanuel Kant, *Critique of Pure Reason* (1787; reprint, London: Macmillan, 1950).

6. Bruno Latour, "The Three Little Dinosaurs" (forthcoming). [This citation could not be updates and was not updated by Latour and Woolgar in the 1986 edition of the book. Ed.]

7. This has been the stock in trade of philosophers since Hume's radical treatment of the problem.

8. Roy Bhaskar, *A Realist Theory of Science* (Atlantic Heights, N.J.: Humanities Press, 1975), p. 10.

9. Ibid., p. 21.

10. Ibid., p. 22.

11. When asked to describe the object of a statement which has been "discovered," scientists invariably repeated the statement. By repeating the same statement in less detail, however, it is possible to convey the impression that there is *more* to reality than is being said. The incompleteness of this description is taken as an indication that the object is not entirely exhausted by knowledge of it. See Jean Paul Sartre, *L'Etre et le Néant* (Paris: Gallimard, 1943).

12. Karl Marx, *Feuerbach: Opposition of the Naturalistic and Idealistic Outlook* (New York: Beckman, 1970).

13. The history of the construction of this substance will be related in detail elsewhere. By contrast with the case of TRF, the observer was present from initial attempts to construct this substance up until its final solidification and use in industrial processes.

14. Latour, "The Three Little Dinosaurs."

15. The question now raised is what kind of explanation is applicable to the settlement of controversy, given that its truth statement cannot be used. Although we indicate some of the answers in the case of TRF, and go on to outline a general model of explanation, our main intention here is to extricate the question from the remnants of the realist position.

16. Our use of the term *shadow* contrasts with Plato's original use of the term. For us, reality (ideas in Plato's terms) is the shadow of scientific practice.

17. Frequently, in histories of epistemology (for example, Gaston Bachelard, *Le Nouvel Espirit Scientifique* [Paris: PUF, 1934]) the argument of efficacy is used when the argument of truth becomes untenable; conventionalists take oven (Henri Poincaré, *Science and Hypothesis* [New York: Dover, 1905]) when realists are defeated (and vice versa). The argument that it works is not more or less mysterious than the argument that it fits reality.

18. Many of the substances (and their analogues) are patented. Substances "discovered" in the laboratory are described in the texts of patents as having been "invented." This shows that the ontological status of statements is rarely likely to be finally settled: depending on the prevailing interests of the parties concerned, the "same" substance can be given a new status.

19. The notion of reliability is itself subject to negotiation (Harry M. Collins, "The T.E.A. Set: Tacit Knowledge and Scientific Networks," *Science Studies* 4 [1974]: 165–96; Collins, "The Seven Sexes"). When, for example, several laboratories failed to confirm the results produced by members of our laboratory, the latter simply recast these failures as evidence of the others' incompetence.

20. We do not wish to argue another version of the induction problem in philosophical terms; we simply want to put the problem on an empirical footing so as to make it amenable to study by sociologists of science. On an empirical basis, neither TRF nor somatostatin escapes the material and social networks in which they are continually constructed and deconstructed. For a discussion of the case of somatostatin, see Paul Brazeau and Roger Guillemin, "Somatostatin: Newcomer from the Hypothalamus," *New England Journal of Medicine* 290 (1974): 963–64.

21. Baruch du Spinoza, *The Ethics,* appendix pt. 1 (1677; reprint, Seacaucus, N.J.: Citadel Press, 1976).

22. This wonderment is particularly marked in matters of science. Nobody wonders that the first steam engine from Newcastle has now

developed into a worldwide railway network. Similarly, nobody takes this extension as the proof that a steam engine can circulate even where there are no rails! By the same token, it has to be remembered that the extension of a network is an expensive operation, and that steam engines circulate only on the lines upon which it has been made to circulate. Even so, observers of science frequently marvel at the "verification" of a fact within a network in which it was constructed. At the same time they happily forget the cost of the extension of the network. The only explanation for this double standard is that a fact is supposed to be an idea. Unfortunately, empirical observation of laboratories makes this idealization of facts impossible.

TWO

CONCLUSION TO *LEVIATHAN AND THE AIR-PUMP*

STEVEN SHAPIN AND SIMON SCHAFFER

Solutions to the problem of knowledge are solutions to the problem of social order. That is why the materials [presented here] are contributions to political history as well as to the history of science and philosophy. Hobbes and Boyle proposed radically different solutions to the question of what was to count as knowledge: which propositions were to be accounted meaningful and which absurd, which prob-

---

lems were soluble and which not, how various grades of certainty were to be distributed among intellectual items, where the boundaries of authentic knowledge were to be drawn. In so doing, Hobbes and Boyle delineated the nature of the philosophical life, the ways in which it was permissible or obligatory for philosophers to deal with each other, what they were to question and what to take for granted, how their activities were to relate to proceedings in the wider society. In the course of offering solutions to the question of what proper philosophical knowledge was and how it was to be achieved, Hobbes and Boyle specified the rules and conventions of differing philosophical forms of life. We conclude by developing some ideas about the relationships between knowledge and political organization.

There are three senses in which we want to say that the history of science occupies the same terrain as the history of politics. First, scientific practitioners have created, selected, and maintained a polity within which they operate and make their intellectual product; second, the intellectual product made within that polity has become an element in political activity in the state; third, there is a conditional relationship between the nature of the polity occupied by scientific intellectuals and the nature of the wider polity. We can elaborate each of these points by refining a notion we have used informally throughout this book: that of an intellectual *space*.[1]

Our previous usages of terminology such as "experimental space" or "philosophical space" have been twofold: we have referred to space in an abstract sense, as a cultural domain. This is the sense customarily intended when one speaks of the boundaries of disciplines or the overlap between areas of culture. The cartographic metaphor is a good one: it reminds us that there are, indeed, abstract cultural boundaries that exist in social space. Sanctions can be enforced by community members if the boundaries are transgressed. But we have also, at times, used the notion of space in a physically more concrete sense. The receiver of the air-pump circumscribed such a space, and we have shown the importance attached by Boyle to defending the integrity of that space. Yet we want to elaborate some notions concerning a rather larger-scale physical space. If someone were to be asked in 1660, "Where can I find a natural philosopher at work?" to what place would he be directed? For Hobbes there was to be no special space in which one did natural philosophy. Clearly, there were spaces that were deemed grossly inappropriate. Since philosophy was a noble activity, it was not to be done in the apothecary shop, in the garden, or in the tool room. He told his adversaries that philosophers were not "apothecaries," "gardeners," or any other sort of "workmen." Neither was philosophy to be withdrawn into the Inns of Court, the physicians' colleges, the clerics' convocations, or the universities. Philosophy was not the exclusive domain of the professional man. Any such withdrawal into special professional spaces threatened the public status of philosophy. Recall Hobbes's indictment of the Royal Society as yet another restricted professional space. He asked, "Cannot anyone who wishes come?" and gave the answer, "The place where they meet is not public."[2] We have seen that the experimentalists also insisted upon the public nature of their activity, but Boyle's "public" and Hobbes's "public" were different usages. Hobbes's philosophy had to be public in the sense that it must not become the preserve of interested professionals. The special interests of professional groups had acted historically to corrupt knowledge. Geometry had escaped this appropriation only because, as a contingent historical matter, its theorems and findings had not been seen to have a bearing on such interests: "Because men care not, in that subject, what be truth, as a thing that crosses no man's ambition, profit or lust."[3] Hobbes's philosophy also had to be public because its purpose was the establishment of public peace and because it commenced with social acts of agreement: settling the meanings and proper uses of words. Its public was not a witnessing and believing public, but an assenting and professing public: not a public of eyes and hands, but one of minds and tongues.

In Boyle's program there was to be a special space in which experimental natural philosophy was done, in which experiments were performed and witnessed. This was the nascent *laboratory*. . . .

What little we do know about English experimental spaces in

the middle part of the seventeenth century indicates that their status as private or public was intensely debated. The word "laboratory" arrived in English usage in the seventeenth century, carrying with it apparently hermetical overtones: the space so designated was private, inhabited by "secretists." During the 1650s and 1660s new open laboratories were developed, alongside Boyle's rhetorical efforts to lure the alchemists into public space and his assaults on the legitimacy of private practice. The public space insisted upon by experimental philosophers was a space for collective witnessing. We have shown the importance of witnessing for the constitution of the matter of fact. Witnessing was regarded as effective if two general conditions could be satisfied: first, the witnessing experience had to be made accessible; second, witnesses had to be reliable and their testimony had to be creditable. The first condition worked to open up experimental space, while the second acted to restrict entry. What in fact resulted was, so to speak, a public space with restricted access. (Arguably, this is an adequate characterization of the scientific laboratory of the late twentieth century: many laboratories have no legal sanction against public entry, but they are, as a practical matter, open only to "authorized personnel.") Restriction of access, we have indicated, was one of the positive recommendations of this new experimental space in Restoration culture. Either by decision or by tacit processes, the space was restricted to those who gave their assent to the legitimacy of the game being played within its confines.

[Earlier] we described differences in the engagements Boyle conducted with two sorts of adversaries: those who disputed moves within the experimental game and those who disputed the game. The latter could be permitted entry to the experimental community only at the price of putting that community's life at risk. Public stipulations about the accessibility of the experimental laboratory were tempered by the practical necessity of disciplining the experimental collective. This tension meant that Hobbes's identification of the Royal Society as a restricted place was potentially damaging, just as it is damaging in modern liberal societies to remark upon the sequestration of science. Democratic ideals and the exigencies of professional expertise form an unstable compound.[4] Hobbes's identification of restrictions on the experimental public shows why virtual witnessing was so vitally important, and why troubles in the experimental program of physical replication were so energetically dealt with. Virtual witnessing acted to ensure that witnesses to matters of fact could effectively be mobilized in abstract space, while securing adequate policing of the physical space occupied by local experimental communities.

For Hobbes, the activity of the philosopher was not bounded: there was no cultural space where knowledge could be had where the philosopher should not go.[5] The methods of the natural philosopher were, in crucial respects, identical to those of the civic philosopher, just as the purpose of each was the same: the achievement and protection of public peace. Hobbes's own career was a token of the philosophical enterprise so conceived. For Boyle and his colleagues, the topography of culture looked different. Their cultural terrain was vividly marked out with boundary-stones and warning notices. Most importantly, the experimental study of nature was to be visibly withdrawn from "humane affairs." The experimentalists were not to "meddle with" affairs of "church and state." The study of nature occupied a quite different space from the study of men and their affairs: objects and subjects would not and could not be treated as part of the same philosophical enterprise. By erecting such boundaries, the experimentalists thought to create a quiet and a moral space for the natural philosopher: "civil war" within their ranks would be avoided by observing these boundaries and the conventions of discourse within them. They would not speak of that which could not be mobilized into a matter of fact by the conventionally agreed patterns of community activity—thus the importance of legislation against speech about entities that would not be made sensible: either those that indisputably *did* exist (e.g., God and immaterial spirits) or those that probably did not (e.g., the aether). As a practical matter, Hobbes could hardly deny that the experimentalists had established a community with some politically important characteristics: a community whose members endeavored to avoid metaphysical talk and causal inquiry, and which displayed many of the attributes of internal peace. But this community was not a society of *philosophers.* In abandoning the philosophical quest, such a group was contributing to civil disorder. It was the philosopher's task to secure public peace; this he could only do by rejecting the boundaries the experimentalists proposed between the study of nature and the study of men and their affairs.

The politics that regulated transactions between the philosophical community and the state was important, for it acted to characterize and to protect the knowledge the philosopher produced. The politics that regulated transactions within the philosophical community was equally important, for it laid down the rules by which authentic knowledge was to be produced. Hobbes assumed philosophical places to have "masters": Father Mersenne had been such a master in Paris, and Hobbes spoke of Boyle and some few of his friends "as masters of the rest" in the Royal Society. It was fitting that philosophical places should have masters who determined right philosophy, just as it was right and necessary that the commonwealth should have such a master. Indeed, Leviathan could legitimately act as a philosophical master. Hobbes found it no argument against the King's right to determine religious principles that "priests were better instructed," and he also rejected the argument "that the authority of teaching *geometry* must not depend upon kings, except they themselves were geometricians."[6] Insofar as a philosophical master was not Leviathan, he was someone else who had found out fundamental matters: the correct principles upon which a unified philosophical enterprise could proceed. He was a master by virtue of his exercise of pure mind, not by his craft-skills or ingenuity. In the body politic of the Hobbesian philosophical place, the mind was the undisputed master of the eyes and the hands.

In the body politic of the experimental community, mastery was *constitutionally restricted.* We have seen how Hooke described the experimental body in terms of the relationships that ought to subsist between intellectual faculties: "The *Understanding* is to *order* all the inferior services of the lower Faculties; but yet it is to do this only as a *lawful Master,* and not as a *Tyrant.*" The experimental polity was an organic community in which each element crucially depended upon all others, a community that rejected absolute hierarchical control by a master. Hooke continued:

> So many are the *links,* upon which the true Philosophy depends, of which, if any one be *loose,* or *weak,* the whole *chain* is in danger of being dissolv'd; it is to *begin* with the Hands and Eyes, and to *proceed* on through the Memory, to be *continued* by the Reason; nor is it to stop there, but to *come about* to the Hands and Eyes again, and so, by a *continual passage round* from one Faculty to another, it is to be maintained in life and strength, as much as the body of man is.[7]

The experimental polity was said to be composed of free men, freely acting, faithfully delivering what they witnessed and sincerely believed to be the case. It was a community whose freedom was responsibly used and which publicly displayed its capacity for self-discipline. Such freedom was safe. Even disputes within the community could be pointed to as models for innocuous and managed conflict. Moreover, such free action was said to be requisite for the production and protection of objective knowledge. Interfere with this form of life and you will interfere with the capacity of knowledge to mirror reality. Mastery, authority, and the exercise of arbitrary power all acted to distort legitimate philosophical knowledge. By contrast, Hobbes proposed that philosophers should have masters who enforced peace among them and who laid down the principles of their activity. Such mastery did not corrode philosophical authenticity. The Hobbesian form of life was not, after all, predicated upon a model of men as free-acting, witnessing, and believing individuals. Hobbesian man differed from Boylean man precisely in the latter's possession of free will and in the role of that will in constituting knowledge. Hobbesian philosophy did not seek the foundations of knowledge in witnessed and testified matters of fact: one did not ground philosophy in "dreams." We see that both games proposed for natural philosophers assumed a causal connection between the political structure of the philosophical community and the genuineness of the knowledge produced.

Hobbes's philosophical truth was to be generated and sustained by absolutism. Boyle and his colleagues lacked a precise vocabulary for the polity they were attempting to erect. Almost all of the terms they used were highly contested in the early Restoration: "civil society," a "balance of powers," a "commonwealth." The experimental community was to be neither tyranny nor democracy. The "middle wayes" were to be taken.[8]

Scientific activity, the scientist's role, and the scientific community have always been dependent: they exist, are valued, and supported insofar as the state or its various agencies see point in them. What sustained the experimental space that was created in the mid-seventeenth century? The nascent laboratory of the Royal Society and other experimental spaces were producing things that were widely wanted in Restoration society. These wants did not simply preexist, waiting to be met; they were actively cultivated by the experimentalists. The experimentalists' task was to show others that their problems could be solved if they came to the experimental philosopher and to the space he occupied in Restoration culture.[9] If the experimentalists could effectively cultivate and satisfy these wants, the legitimacy of experimental activity and the integrity of laboratory and scientific role would be ensured. The wants addressed by the experimental community spread across Restoration economic, political, religious, and cultural activity. Did gunners want their artillery pieces to fire more accurately? Then they should bring their practical problems to the physicists of the Royal Society. Did brewers want a more reliable ale? Then they should come to the chemists. Did physicians want a theoretical framework for the explanation and treatment of fever? Then they should inspect the wares of the mechanical philosopher. The experimental laboratory was advertised as a place where practically useful knowledge was produced.[10] But the laboratory could also supply solutions to less tangible problems. Did theologians desire facts and schemata that could be deployed to convince otherwise obdurate men of the existence and attributes of the Deity? They, too, should come to the laboratory where their wants would be satisfied. Through the eighteenth century one of the most important justifications for the natural philosopher's role was the spectacular display of God's power in nature.[11] Theologians could come to the place where the Leyden jar operated if they wanted to show cynics the reality of God's majesty; natural theologians could come to the astronomer's observatory if they wanted evidence of God's wise and regular arrangements for the order of nature; moralists could come to the natural historian if they wanted socially usable patterns of natural hierarchy, order, and the due submission of ranks. The scientific role could be institutionalized and the scientific community could be legitimized insofar as the experimental space became a place where this multiplicity of interests was addressed, acquitted, and drawn together. One of the more remarkable features of the early experimental program was the intensity with which its proponents worked to publicize experimental spaces as useful: to identify problems in Restoration society to which the work of the experimental philosopher could provide the solutions.

There was another desideratum the experimental community sought to mobilize and satisfy in Restoration society. The experimental philosopher could be made to provide a model of the moral citizen, and the experimental community could be constituted as a model of the ideal polity. Publicists of the early Royal Society stressed that theirs was a community in which free discourse did not breed dispute, scandal, or civil war; a community that aimed at peace and had found out the methods for effectively generating and maintaining consensus; a community without arbitrary authority that had learned to order itself. The experimental philosophers aimed to show those who looked at their community an idealized reflection of the Restoration settlement. Here was a functioning example of how to organize and sustain a peaceable society between the extremes of tyranny and radical individualism. Did civic philosophers and political actors wish to construct such a society? Then they should come to the laboratory to see how it worked.

[We have] been concerned with the identification of alternative philosophical forms of life, with the display of their conventional bases, and with the analysis of what hinged upon the choice between them. We have not taken as one of our questions, "Why did

Boyle win?" Obviously, many aspects of the program he recommended continue to characterize modern scientific activity and philosophies of scientific method. Yet, an unbroken continuum between Boyle's interventions and twentieth-century science is highly unlikely. For example, the relationship between Boyle's experimental program and Newton's "mathematical way" is yet to be fully explored. Nevertheless, modern historians who find in Boyle the "founder" of truly modern science can point to similar sentiments among late-seventeenth-century and eighteenth-century commentators. Despite these qualifications the general form of an answer to the question of Boyle's "success" begins to emerge, and it takes a satisfyingly historical form. This experimental form of life achieved local success to the extent that the Restoration settlement was secured. Indeed, it was one of the important elements in that security.

Insofar as we have displayed the political status of solutions to problems of knowledge, we have not referred to politics as something that happens solely outside of science and which can, so to speak, press in upon it. The experimental community vigorously developed and deployed such boundary-speech, and we have sought to situate this speech historically and to explain why these conventionalized ways of talking developed. What we cannot do if we want to be serious about the historical nature of our inquiry is to use such actors' speech unthinkingly as an explanatory resource. The language that transports politics outside of science is precisely what we need to understand and explain. We find ourselves standing against much current sentiment in the history of science that holds that we should have less talk of the "insides" and "outsides" of science, that we have transcended such outmoded categories. Far from it; we have not yet begun to understand the issues involved. We still need to understand how such boundary-conventions developed: how, as a matter of historical record, scientific actors allocated items with respect to *their* boundaries (not ours), and how, as a matter of record, they behaved with respect to the items thus allocated. Nor should we take any one system of boundaries as belonging self-evidently to the thing that is called "science."

We have had three things to connect: (1) the polity of the intellectual community; (2) the solution to the practical problem of making and justifying knowledge; and (3) the polity of the wider society. We have made three connections: we have attempted to show (1) that the solution to the problem of knowledge is political; it is predicated upon laying down rules and conventions of relations between men in the intellectual polity; (2) that the knowledge thus produced and authenticated becomes an element in political action in the wider polity; it is impossible that we should come to understand the nature of political action in the state without referring to the products of the intellectual polity; (3) that the contest among alternative forms of life and their characteristic forms of intellectual product depends upon the political success of the various candidates in insinuating themselves into the activities of other institutions and other interest groups. He who has the most, and the most powerful, allies wins.

We have sought to establish that what the Restoration polity and experimental science had in common was a form of life. The practices involved in the generation and justification of proper knowledge were part of the settlement and protection of a certain kind of social order. Other intellectual practices were condemned and rejected because they were judged inappropriate or dangerous to the polity that emerged in the Restoration. It is, of course, far from original to notice an intimate and an important relationship between the form of life of experimental natural science and the political forms of liberal and pluralistic societies. During the Second World War, when liberal society in the West was undergoing its most virulent challenge, that perception was formed into part of the problematic of the academic study of science. What sort of society is able to sustain legitimate and authentic science? And what contribution does scientific knowledge make to the maintenance of liberal society?[12] The answer then given was unambiguous: an open and liberal society was the natural habitat of science, taken as the quest for objective knowledge. Such knowledge, in turn, constituted one of the sureties for the continuance of open and liberal society. Interfere with the one, and you will erode the other.

Now we live in a less certain age. We are no longer so sure that traditional characterizations of how science proceeds adequately describe its reality, just as we have come increasingly to doubt whether liberal rhetoric corresponds to the real nature of the society in which we now live. Our present-day problems of defining our knowledge, our society, and the relationships between them center on the same dichotomies between the public and the private, between authority and expertise, that structured the disputes we have examined. We regard our scientific knowledge as open and accessible in principle, but the public does not understand it. Scientific journals are in our public libraries, but they are written in a language alien to the citizenry. We say that our laboratories constitute some of our most open professional spaces, yet the public does not enter them. Our society is said to be democratic, but the public cannot call to account what they cannot comprehend. A form of knowledge that is the most open in principle has become the most closed in practice. To entertain these doubts about our science is to question the constitution of our society. It is no wonder that scientific knowledge is so difficult to hold up to scrutiny.

We have examined the origins of a relationship between our knowledge and our polity that has, in its fundamentals, lasted for three centuries. The past offers resources for understanding the present, but not, we think, for foretelling the future. Nevertheless, we can venture one prediction as highly probable. The form of life in which we make our scientific knowledge will stand or fall with the way we order our affairs in the state.

We have written about a period in which the nature of knowledge, the nature of the polity, and the nature of the relationships between them were matters for wide-ranging and practical debate. A new social order emerged together with the rejection of an old intellectual order. In the late twentieth century that settlement is, in turn, being called into serious question. Neither our scientific knowledge, nor the constitution of our society, nor traditional statements about the connections between our society and our knowledge are taken for granted any longer. As we come to recognize the conventional and artifactual status of our forms of knowing, we put ourselves in a position to realize that it is ourselves and not reality that is responsible for what we know. Knowledge, as much as the state, is the product of human actions. Hobbes was right.

NOTES

1. We are not aware of any specific debts for this usage. However, topographic sensibilities in the study of culture characterize a number of modern French sociologists and historians; see, for example, Michel Foucault, "Questions on Geography," in *Power-Knowledge: Selected Interviews and Other Writings,* ed. Colin Gordon (Brighton: Harvester, 1980), pp. 63–77; Foucault, "Médicins, juges et sorciers au 17e siécle," *Médicine ae France* 200 (1969): 12–128.

2. Thomas Hobbes, "Dialogus physicus," appendix to *Leviathan and the Air-Pump,* trans. Simon Schaffer (Princeton: Princeton University Press, 1985).

3. Thomas Hobbes, "Leviathan," in *The English Works of Thomas Hobbes of Malmesbury,* ed. William Molesworth, 11 vols. (London: John Bohn, 1839–45), 3: 91. Hobbes made no claim of the sort that geometry is essentially neutral.

4. This has often been noted by historians dealing with widely differing settings: see, for example, George H. Daniels, "The Pure-Science Ideal and Democratic Culture," *Science* 156 (1967): 1699–1705; Yaron Ezrahi, "Science and the Problem of Authority in Democracy," in *Science and Society: A Festschift for Robert K. Merton,* ed. Thomas F. Gieryn, Transactions of the New York Academy of Sciences, series 2, vol. 39 (New York: New York Academy of Sciences, 1980), pp. 43–60; Sylvia Fries, "The Ideology of Science during the Nixon Years," in *Social Studies of Science,* 14 (1984): 323–41; Charles C. Gillispie, "The Encyclopedié and the Jacobin Philosophy of Science," in *Critical Problems in the History of Science,* ed. Marshall Clagett (Madison: University of Wisconsin Press, 1959), pp. 255–89.

5. According to Hobbes, men "cannot have any idea of [God] in their mind, answerable to his nature" ("Leviathan," p. 92), and, for that reason, theology was explicitly excluded from the philosophical enterprise ("Concerning Body," p. 10).

6. Thomas Hobbes, "Philosophical Rudiments," in *The English Works of Thomas Hobbes of Malmesbury,* ed. William Molesworth, 2: 247.

7. Robert Hooke, preface to *Micrographia*

8. The phrase is Hooke's from *Micrographia.* Similar locutions typify much Royal Society publicity.

9. For this section we are deeply indebted to recent work by Bruno Latour, especially his "Give Me a Laboratory and I will Raise the World," in *Science Observed: Perspectives on the Social Study of Science,* ed. Karin Knorr-Cetina and Michael Milkay (London: Sage, 1983), pp. 141–70; and *Les microbes: guerre er paix suive de irréductions* (Paris: Editions A. M. Métailié, 1983).

10. From the best modern historical research it now appears that none of the utilitarian promissory notes could be, or were, cashed in the seventeenth century. See Richard S. Westfall, "Hooke, Mechanical Technology, and Scientific Investigation," in *The Uses of Science in the Age of Newton,* ed. John G. Burke (Berkeley: University of California Press, 1983); A. R. Hall, "Gunnery, Science, and the Royal Society," also in *The Uses of Science in the Age of Newton.* If science did not deliver technological utility, it becomes even more important to ask about its other perceived values, including social, political, and religious uses.

11. See particularly Simon Schaffer, "Natural Philosophy," in *The Ferment of Knowledge: Studies in the Historiography of Eighteenth Century Science,* ed. G. S. Rousseau and Roy Porter (Cambridge: Cambridge University Press, 1980), pp. 55–91; Schaffer, "Natural Philosophy and Public Spectacle in the Eighteenth Century," *History of Science* 21 (1983): 1–43.

12. Robert K. Merton, *The Sociology of Science: Theoretical and Empirical Investigations,* chaps. 12–13, ed. Norman W. Storer (London: George Allen & Unwin, 1969); Zilsel, *Die sozialen Ursprünge der neuzeitkichen Wissenschaft,* ed. and trans. Wolfgang Krohn (Frankfort Am Main: Suhrkamp, 1976).

THREE

THE SOCIOLOGY OF KNOWLEDGE

Exposition and Critique

ROBERT KLEE

The sociology of knowledge is an empirical discipline. It will stand or fall on whether its pronouncements are consistent with the relevant data. The relevant data are case studies in the history and/or practice of science. It is no wonder then that social constructivist models of science are heavily dependent—

perhaps even more so than Kuhn himself was—on case studies of scientific practice and method. Bruno Latour and Steve Woolgar are two prominent social constructivists who are specialists in the right sort of case studies. The idea is to do the case studies, not the way a historian would, but the way an anthropologist doing field work would. This is what Latour did when he showed up at a research laboratory housed in the Salk Institute in California, a laboratory specializing in the investigation of hormones that originate in the nervous system. He wanted to observe practitioners working in the lab in the hope that they might follow the social construction of a scientific "fact" from beginning to end. Woolgar later joined Latour to coauthor a study based on Latour's observations. They wrote with a heavy dose of the self-referential plural despite Woolgar's absence from the lab. My own remarks below will respect their preference in this matter by discussing their study as if it was based on a jointly experienced sociological field exercise.

Latour and Woolgar admit to approaching science with an attitude not unlike that with which field anthropologists approach the study of a primitive and alien culture. Science to them is alien in the same way a previously unknown tribal culture is alien. Further, just as no well-trained anthropologist would take the primitive natives' own interpretation of their culture as a correct account of it, so Latour and Woolgar argue that it is not appropriate to take scientists' own interpretations of their work as correct accounts of it. The sociologist of science must avoid "going native," as they put it. On the contrary, science is to be accorded initially no more rationality as a knowledge-gathering method than is, say, primitive witchcraft. Latour and Woolgar come close to confessing outright this sort of debunking view of science as witchcraft in a footnote: "One particularly useful source [for their own study] was Auge's . . . analysis of witchcraft in the Ivory Coast, which provides an intellectual framework for resistance to being impressed by scientific endeavour." (*Laboratory Life*, 260). Just as the field anthropologist never really believes that the tribal sorcerer under study is actually breaking the laws of nature with his or her witchcraft, so, according to Latour and Woolgar, the sociologist of science ought never to believe the scientists' own claims that science finds out "the facts," the "actual way things really are." As Latour and Woolgar see it, the task of the sociologist of science is to give a debunking account of scientific practice, one that shows how practitioners of science become deluded into thinking that they are discovering a reality that was already in existence rather than constructing a reality by artificial means (which is actually what they are doing instead). It is no accident therefore when Latour and Woolgar admit unashamedly that they studied the behavior of the scientists in the lab as though those scientists were the postindustrial equivalent of, and here I want to use the very expression Latour and Woolgar use, *primitive sorcerers* (*Laboratory Life*, 29) engaged in routinized behavior that was entirely "strange" and virtually incomprehensible to Latour and Woolgar. One bizarre oddity to their study (which the intelligent reader can hardly miss) is how "strange" Latour and Woolgar seem in failing to understand or appreciate the obvious explanation for this strangeness they felt toward what they were observing in the lab: Namely, unfamiliarity and lack of experience with *any* kind of activity will tend to make that activity seem strange when seen from the outside. But instead of condemning their own ignorance as scientifically naive observers, Latour and Woolgar conclude that they are the only people in the lab capable of an objective understanding of what the scientists are doing. Ignorance is a necessary condition of wisdom, apparently, in the ironic world of the sociology of science.

What those scientists are doing is not what those scientists think that they are doing, according to Latour and Woolgar. The field study is rather humorous to read, although some of the humor is surely unintentional on Latour and Woolgar's part. The two field anthropologists are thoroughly perplexed by the daily activities of the lab workers, activities that the sociologists describe in the most observationally literal way they can invent. Machines hum and churn and produce "wavy lines," which are then superimposed; flasks of liquid are shaken, then argued over; workers consult texts in order to produce more "inscriptions." The attempt is to describe moment-to-moment practice in a bioscience laboratory in such a manner as to make it out to be eccentric and arbitrary enough to qualify as irrational: science as a more fancy form of witchcraft. What truly shocks Latour and Woolgar is the claim made by the practitioners they are studying that they, the practitioners, are doing something much more worthwhile than arbitrary witchcraft—that they are discovering objective structures in the domain of nature under their investigation, that they are finding out objective facts about neurohormones. Latour and Woolgar are sure that this is not so, that any claim to have discovered preexisting material realities is a self-congratulatory form of exaggeration on the part of the scientists, basically nothing more than arrogant hubris; for all that is *really* going on in the lab is that, machines hum and churn and produce wavy lines, flasks of colored liquid are shaken, then argued about, workers consult texts in order to produce more inscriptions, and so on.

The problem seems to be that Latour and Woolgar make a basic mistake in the epistemology of science. They take the fact that, often in science, the detection of an objectively existing structural reality requires an elaborate detection procedure (involving elaborate machines and so on) to mean that no such objectively existing structural realities exist; for what the practitioners are "detecting" in such cases is an artificial structural reality, one created by the elaborate detection procedure. In other words, for Latour and Woolgar, the wavy lines produced by the fancy machine that produces wavy lines are not wavy lines *about anything except the very process that produced them.* The content of science is merely the process by which it is artificially constructed as knowledge—it is not the structure of some independent reality. No properties of an independent reality are represented in the wavy lines, for the wavy lines are entirely artificially manufactured by the machine that makes wavy lines. When the scientists themselves object to this that the wavy lines contain peaks and valleys which represent some of the chemical properties of a hormone sample that was put into the machine that makes wavy lines, Latour and Woolgar will believe none of it; for that is too much like listening to the tribal witch doctors insisting that the spell they cast really did cure the sick daughter of the assistant chief by driving away the independently existing evil spirits.

Latour and Woolgar will have none of it in this way because they make a specific philosophical mistake: From the fact that our *being able to say or to claim* that X exists depends on an elaborate and artificial procedure for detecting X, they mistakenly think it follows that X's existence itself depends on that elaborate and artificial detection procedure. But, of course, this is an absurd as well as an invalid inference. To see the absurdity, consider the following implication of the Latour and Woolgar view. We cannot observe viruses without very elaborate and wholly artificial equipment such as electron microscopes, special chemical stains, and so on. If Latour and Woolgar are correct, that would imply that the *existence* of viruses *depends* on such artificial equipment and such elaborate detection procedures. If such procedures and equipment did not exist, *then viruses would not exist.*

Why do Latour and Woolgar conceive the sociology of science to be this kind of debunking project? What motivates their confidence that practitioners of modern science are as serenely mistaken about the real nature of their own business as are the tribal witch doctors about the real nature of their business? Well, Latour and Woolgar claim that the facts—indeed, all the facts—of science are entirely socially constructed. Scientific facts are really about their own social construction, not about some independent reality. But they realize that everybody thinks scientific facts are about some inquiry-independent reality. Now most scientists are fairly smart. What would lead smart people to make this sort of mistake about the real status of scientific facts? One central chore of the sociology of science must be to explain how it is that what is entirely socially constructed and about one sort of subject matter is almost universally mistaken by otherwise bright human beings to be not entirely socially constructed and about a completely different kind of subject matter. This will force the sociology of science to become primarily a project of debunking. Latour and Woolgar attempt to provide the missing explanation of the mistake: The methods of science are designed to "cover up" the social construction of the products of science. Once a purported fact has been established by scientific means, the elaborate and artificial procedures used to establish it conveniently disappear from notice and are forgotten. The fact now seems to "stand on its own"; hence, the delusion (Latour and Woolgar's word for it is "illusion," but they are being sloppy here—they mean to speak of what is more properly called a delusion) naturally arises that the fact is about an independently existing objective reality and not about its own process of construction through organized social behavior. That, according to Latour and Woolgar, is why the wavy lines produced by the machine that makes wavy lines are universally held to have been made by the interaction of an independent reality with the machine, when actually they were complete creations of the machine and how it was manipulated by its users.

How does the "cover up" occur? Here Latour and Woolgar become very unclear; for, despite their lack of empathy with philosophy, they use a central philosophical notion in their account of the covering up process: the notion of a modality. In modal logic and metaphysics philosophers and logicians have worked out relatively precise characterizations of various modalities. Unfortunately, Latour and Woolgar are completely innocent of this prior work. They don't quite mean by "modality" what philosophers and modal logicians mean by "modality." Just what they do mean is not entirely clear. A modality is a "mode" that qualifies the truth-conditions of a sentence. For example, "possibly" is taken to set up a modal context. So, the sentence,

UFOs are spacecraft piloted by alien beings

does not mean the same (because it is not true or false under *exactly* the same conditions) as the sentence,

Possibly, UFOs are spacecraft piloted by alien beings.

Even if we knew that the first sentence is false, the second one could still be true—that's the difference that "possibly" makes to the truth-conditions for what would be otherwise the same sentence. There are many kinds of nonmodal sentences that nonetheless function logically like modalities that the philosopher Bertrand Russell called *propositional attitudes*. Propositional attitudes have to do with the different psychological attitudes that a human being can take "toward" a sentence (actually, the attitude is taken toward the content of the sentence, which philosophers call the proposition of the sentence). For example, the sentence,

Some UFOs are spacecraft piloted by alien beings

does not mean the same as, indeed, it has very different truth-conditions, than the sentence,

Harvey is skeptical that some UFOs are spacecraft piloted by alien beings.

And both those sentences are yet different in meaning from the sentence,

Harvey is confident that some UFOs are spacecraft piloted by alien beings;

which is different again from the sentence,

Harvey wishes it weren't commonly believed that some UFOs are spacecraft piloted by alien beings.

I have put all the different propositional attitude indicator phrases in italics in the above examples. The reader can see the basic pattern. The same declarative sentence occurs after a different attitude phrase ending with the word "that." This is the commonly accepted way of representing propositional attitudes. Notice that the truth-value of the buried declarative sentence can stay the same while each sentence as a whole can change truth-value depending on the content of the attitude phrase. Suppose, for example, that as a matter of fact it is false that some UFOs are spacecraft piloted by alien beings. Nevertheless, it can be true that Harvey is skeptical of this fact, false that he is confident of that fact, and true that he wishes this falsehood weren't so commonly believed to be true by so many other people. The reader can now sense the virtually limitless number of different propositional attitudes: "believes that . . . ," "fears that . . . ," "knows that . . . ," "doubts that . . . ," and so on, are all propositional attitudes in which a declarative sentence fills the ". . ." position. Here is the important point about propositional attitudes for purposes of our present discussion: The truth or falsehood of a propositional attitude sentence is not purely a function of the truth or falsehood of the embedded declarative sentence—rather, the specific content of the attitude factors is also.

Latour and Woolgar are misled by the presence of propositional attitudes in scientific reasoning and practice into the belief that all scientific facts are about their own construction, not about some practice-independent reality. What misleads them is how different propositional attitudes occur at different times in the process of discovering a scientific fact. The attitudes expressed early in the discovery process tend to be attitudes involving skepticism and uncertainty. Surely this ought to be no surprise. Equally, it ought to be no surprise, that if scientists are fortunate and their conjectures are true, nature rewards their researches with success, and they begin to express attitudes toward the same "fact" that involve increasing confidence and sureness. For example, a research paper appearing in an immunology journal in 1987 might contain the following propositional attitude sentence,

We now suspect that NK cells kill some of their targets with the aid of antibodies.

This sentence indicates a tentative attitude toward the alleged fact in question. It is being surmised, floated as a possibility that subsequent practice will either support or reject. We can imagine it happening that by 1998 the same embedded sentence appears in a research paper attached to a more confident attitude,

We now believe beyond doubt that NK cells kill some of their targets with the aid of antibodies.

Latour and Woolgar take this evolution of attitudes from uncertainty to confidence to show that the belief scientists have that they are discovering preexisting material realities is a delusion; for, according to Latour and Woolgar, the move to more confident attitudes is a purely social creation, one having to do with how many times a given paper is cited in other papers, in textbooks, and in conversation, how effectively the prestige of one scientist can intimidate others into accepting his or her views, how standardized lab machinery becomes so that the same "wavy lines" are reproducible in different labs, and so on, and not one having to do with uncovering a fixed material reality. Their argument is instructively erroneous enough to be worth quoting at length:

> In simple terms, [scientists] were more convinced that an inscription unambiguously related to a substance "out there," if a similar inscription could also be found. In the same way, an important factor in the acceptance of a statement was the recognition by others of another statement that was similar. The combination of two or more apparently similar statements concretised the existence of some external object or objective condition of which statements were taken to be indicators. Sources of "subjectivity" thus disappeared in the face of more than one statement, and the initial statement could be taken at face value without qualification. . . . An "object" was thus achieved through the superimposition of several statements or documents in such a way that all the statements were seen to relate to something outside of, or beyond, the reader's or author's subjectivity (*Laboratory Life*, 83–84).

External objects are said to be "concretised" inventions due to the repeatability of a given "statement" in the research literature. Objects are said to be "achieved" as constructions out of socially produced data that are repeatedly obtainable only because of the standardization of research equipment and procedures. It is no wonder then that later in their study Latour and Woolgar call the objectivity of the concretised, achieved material reality a mere "impression of objectivity" (*Laboratory Life*, 90). It isn't real objectivity at all, is the clear implication. The scientists are as deluded as the tribal witch doctors. If this isn't antirealism by the case method with a vengeance, one wonders what would be.

It is important for us to understand what has happened to the allegedly objective world in the hands of Latour and Woolgar. Here again, we see at work the pernicious influence of the Kantian distinction between phenomena and noumena. *The noumenal world is taken to be so epistemologically remote by Latour and Woolgar as to be never represented in the output of our detecting devices at any stage of our detecting procedures.* Put a slightly different way that makes clearer the antirealism, *the amount of "filtration" through artificial manipulation that scientific data undergo during collection is so great as to destroy whatever was represented about the objective world in the original input at the distal end of the data collection process.* We might call this "The Great Filtration" or "The Great Distortion." It has a distinctly Kantian flavor to it. If one seriously believes in The Great Filtration/Distortion, then it will follow as a matter of course that science is about its own process of construction rather than about an objective world. Again, I quote Latour and Woolgar at length:

> Despite the fact that our scientists held the belief that the inscriptions could be representations or indicators of some entity with an independent existence "out there," we have argued that such entities were constituted solely through the use of these inscriptions. It is not simply that differences between curves [on superimposed graphs] indicate the presence of a substance; rather the substance is identical with perceived differences between curves. In order to stress this point, we have eschewed the use of expressions such as "the substance was discovered by using a bioassay" or "the object was found as a result of identifying differences between two peaks." To employ such expressions would be to convey the misleading impression that the presence of certain objects was a pregiven and that such objects merely awaited the timely revelation of their existence by scientists. By contrast, we

> do not conceive of scientists using various strategies as pulling back the curtain on pregiven, but hitherto concealed, truths. Rather, objects (in this case, substances) are constituted through the artful creativity of scientists (*Laboratory Life*, 128–29).

One hardly knows what to make of the claim that a substance is "identical" with perceived differences between superimposed graphed curves. Bloor promised that the sociology of science would not be a form of idealism, but Latour and Woolgar in the above quote would seem to be wavering on the precipice of idealism. Substances are said to be identical with certain perceptual experiences. One doesn't know whether to laugh or cry when Latour and Woolgar tell the story of TRF (thyrotropin releasing factor, a neurohormone) as though TRF, the actual physical hormone itself (not merely facts *about* it), was created ("a major ontological change took place" *Laboratory Life*, 147); "a decisive metamorphosis occurred in the nature of the constructed object" (*Laboratory Life*, 148) sometime between 1962 and 1969 in two particular laboratories.

Latour and Woolgar claim that the upshot of their particular brand of social constructivism is that *reality is the consequence of the content of science, not its cause.* Because they hold this view, they object to the idea that one could explain why different scientists arrive at similar or convergent beliefs by appealing to an independent world whose stable structure is the same for all scientists; rather, Latour and Woolgar insist, a common independent world is the entirely constructed outcome of the practices of scientists. It follows that realism gets it backward: The structure of the world doesn't constrain the content of science, the content of science constrains the structure of the world. The realist philosopher of science Philip Kitcher argues that for Latour and Woolgar to make good on this last claim they would have to show that differences in scientists' "encounters with the world" (read, with a noumenal, independent material reality) would make no difference to the outcome of what counts as the scientific facts, that no matter how those encounters with the world had been different, scientific facts would have been constructed with the same content. This, Kitcher argues, the two sociologists of knowledge fail to show with their case studies. I might add that what those case studies do show is that differences in the constructed content of scientific facts make a difference to the quality of our encounters with the world. But one could find enlightened realists who would agree with that. If Kitcher is correct, then Latour and Woolgar's interpretation of their study done at the Salk Institute is underargued and a rather misleading exaggeration.

In Latour and Woolgar's view, far from science being constrained in its content by a preexisting world "out there" whose structure it slowly discovers, science is a self-justifying collection of arcane technological practices—replete with mystifying jargon, guild-like exclusivity of participation, and all the trappings of ideological self-centeredness—a collection of arcane technological practices that literally invents the world from scratch. This is nuclear-strength antirealism if anything is.

Latour and Woolgar fess up to their antirealism eventually when they get around to a brief discussion of who they regard as the opposition. They identify the enemy of the sociology of knowledge to be what they call "a realist theory of science" (*Laboratory Life*, 178). Their argument against realism is so woolly as to be nearly unrecoverable, but their basic idea is that realism is a delusion. Their main reason for holding that realism is a delusion seems to be the old howler of a view that every philosophy teacher since the formal teaching of philosophy began has encountered coming out of the mouths of novice students in introductory-level philosophy courses: Namely, because you can't *talk about* an objective and independent material world without using concepts formed during the socially determined investigation of such a world, there is no such world at all—there are only the socially determined practices of investigation that produce the said concepts and the talk we make with them. The mistaken reasoning here is blatant and irreparable.

Latour and Woolgar's argument: If there were an independent material world, it would have to be directly knowable through nonlinguistic, nonconceptual means. But the second we start to talk about what we claim to know of such a world, we are helping ourselves to socially constructed concepts; and, if the concepts are socially constructed, so also are what they are concepts about.

The last clause in this argument is the sort of mistake that ought to remind the reader of one of Bishop Berkeley's more embarrassing moments: Because you can't think that trees exist independently of being thought about without, like, *thinking* about them, there are no such trees—there are only thoughts about trees; or rather, trees "just are" nothing but thoughts about trees. The way out from under this woolly argument of Latour and Woolgar's is surely pretty obvious: The first sentence of the argument is just plain false. Who says that if there is an independent material world it must be knowable directly by nonlinguistic, nonconceptual means? Certainly no realist would make such an implausible demand. Only the Kantian idealist would think up such a self-defeating and question-begging requirement. But Latour and Woolgar, if they are anything philosophical at all, are true to their Kantian (hence, to their idealist) roots: For them what exists collapses into what is knowable—in good Kantian idealist fashion precisely because for them what exists collapses into what is investigatable, and what is investigatable collapses into what is describable, and what is describable collapses into what is knowable. In this tortuous way, Latour and Woolgar lead the sociology of knowledge up a dead-end alley into full-blooded antirealism, an antirealism that comes perilously close to what Bloor said the sociology of knowledge must avoid: an idealist model of science.

THE VIEW FROM DEEP IN THE SWAMP OF HISTORY

It was inevitable that Kuhn's historicizing of philosophy of science would lead to a host of extremely detailed studies of particular episodes in the history of science. Book-length studies were surely destined to appear in which scholars would meticulously pour over the small-scale historical trivia of science: Galileo's private letters to his mistresses, Einstein's eating habits between thought-experiments, Darwin's sexual inhibitions as revealed by remarks in his journal on the breeding of domestic animals. I do not know whether any treatises have been published on the three above topics, but one social constructivist historical study did appear that caused a bit of a stir: Steven Shapin and Simon Schaffer's examination of the dispute between Robert Boyle and Thomas Hobbes over Boyle's air-pump and what it allegedly showed about the role of experimentation in science, a dispute that started in 1660 and lasted until Hobbes's death in 1679. The standard view of this dispute makes Boyle out to be the hero and Hobbes the villain. Boyle's position is taken to be the correct one while Hobbes's position is often portrayed as that of a boorish crank's, as though Hobbes was a mere armchair philosopher (whereas Boyle was a laboratory scientist) pathetically railing against the legitimacy of scientific experimentation from his moldy a priorist perch.

Shapin and Schaffer beg to differ. They seek in their study to show that Boyle didn't know as much as he has been thought to have known, that Hobbes was not quite the boorish crank the standard view says he was, and that Boyle's victory in the dispute was not due to any genuine superiority of his position vis-à-vis the objective facts about the physical world. Hobbes's rejection of Boyle's interpretations of certain experiments with his air-pump was cogent on its own terms, claim Shapin and Schaffer, and the wrongness of Hobbes's position is not much more than partisan history of science as written by the followers of Boyle. That is certainly an ambitious goal to aim their study at, and if Shapin and Schaffer could pull it off, then we should all be grateful for their having rectified a huge mistake in the heretofore accepted understanding of a critically important episode in the history of science.

In the mid-seventeenth century science was not yet understood as an empirical form of inquiry clearly distinguished from traditional philosophy. In fact it was common to refer to what we would now call science by the label "natural philosophy." As an example of how mingled together science and philosophy were, the philosopher Thomas Hobbes in his main work published in 1655 *defined* philosophy as the study of the causes of things. No philosopher nowadays would find that definition plausible at all: Philosophy is not a causal enterprise, it is an analytical form of inquiry, one that seeks to clarify, systematize, and criticize the concepts and methods of other more specialized domains of inquiry. The distinction we now take for granted between issues involving the conventional definitions of words and issues involving matters of fact about physical reality was not nearly as clear in 1660, even to the best minds of the times. It was a time when that distinction was first being worked out by human beings.

Robert Boyle was a brilliant natural philosopher who championed a new way of gathering knowledge, a way that did not fit the traditional form knowledge gathering took in established philosophy. Boyle was suspicious of grand a priori systems, of philosophers who presented complicated deductive systems of definitional truths that the structure of the world must supposedly match because reason itself supposedly necessitated such a match. A grand a priori system was precisely what Hobbes had labored to produce over many years and presented to the world in his monumental *De Corpore* in 1655. Against such overbearing systems of metaphysics Boyle advocated something brand new: the experimental matter of fact. Where the former kinds of metaphysical systems were all doubtful because they were entirely speculative, the latter facts of the matter were certain because they were directly witnessed through ordinary sense perception. Boyle made a distinction between knowing what a matter of fact is and knowing what its cause is. The former kind of knowledge is available to inquiring humans who perform the right sorts of experiments, the latter sort of knowledge is largely forbidden humans because it is beyond their godgiven means of perception to ascertain the hidden structures of things. Hobbes was of course aghast at such a distinction, for it meant that properly philosophical knowledge—knowledge of causes—was for the most part inaccessible to human beings. Hobbes was suspicious of experimentation, of mechanically inclined tinkerers who invented an artificial machine and then claimed to have discovered this and that about the world on the basis of the machine's behavior under artificial conditions.

Obviously, bad blood was bound to flow between Hobbes and Boyle. Boyle invented an air-pump, a machine he claimed evacuated almost all the air from an enclosed receptacle. He published a book reporting the results of forty-three experiments performed with the air-pump. This was a watershed moment in the history of science; for with Boyle's air-pump we see the arrival on the scene of two permanent institutional features of modern science: the standardized physical laboratory as a public space for experimentation, and the official skepticism toward all things theoretical (as opposed to all things observational) that became a universal requirement of the scientific mind-set—so universal and so required that Karl Popper later made it the very heart of scientific method with his falsificationism. Boyle's air-pump, as Shapin and Schaffer show, was the "big-money" science of its time, the seventeenth-century equivalent of a modern day subatomic particle accelerator or a space shuttle. So fancy was it, so difficult and expensive to design and produce was it, that no more than four or five working air-pumps ever existed in the entire world during the first decade of the Boyle/Hobbes dispute. By modern standards, of course, making Boyle's air-pump would be a very achievable project for a bright student competing in a school science fair. No matter, Boyle's air-pump was once the center of an intellectual firestorm during the birth pangs of modern science. Boyle claimed that some of his experiments with the air-pump proved certain incontestable "matters of fact": that the air has a "spring" to it, that the air has a weight to it different from its spring, that a vacuum is possible in nature, and so plenism (the philosophical position that a vacuum in nature is impossible on a priori grounds) is false. Hobbes was an "antivacuist," one of the armchair philosophers who held that a vacuum was impossible on grounds of pure reason. Naturally, Hobbes objected to Boyle's claim that the air-pump experiments involved the creation of a vacuum or near-vacuum inside the pumped receptacle. Shapin and Schaffer present the subsequent dispute in all its floridly bitter detail. A host of side characters appear in the lengthy drama, among them the Dutch natural philosopher Christian Huygens, Robert Hooke, a Jesuit priest named Franciscus Linus, and

Henry More. Round and round they bickered and argued, experimented and counterexperimented (except for Hobbes and his fellow philosopher Spinoza who both scoffed at experimentation as inferior to pure reason), and wrote treatises, tracts, and diatribes. Hobbes insisted that Boyle's air-pump leaked—indeed, that it *must* leak on a priori grounds—for a vacuum requires self-moving matter, which is a self-contradiction. Boyle insisted that the air-pump's leakage was negligible, that it created a vacuum for all *practical* purposes, and that the vacuum it created was a certain matter of fact on which all experimentalists could agree because it was publicly witnessable and repeatable. Nothing less than what form of life science would take was at stake, according to Shapin and Schaffer: Would it be the experimental form of life, or the form of life of a priori metaphysical system building? We can know only observational facts for sure, said Boyle, we are restricted to mere speculation about the underlying causes. Knowledge is of causes and how they fit into a grand metaphysical system or it is of nothing, said Hobbes, the so-called observational facts are mere artifacts of the artificial machines used to create them.

THE INEQUALITY OF THEORETICAL ALTERNATIVES

Shapin and Schaffer's study is interesting and important. It shows just how conceptually messy science can be, especially when it was in its infancy and nothing could be taken for granted about it. The two historians take the underdetermination of theory by observational data as an established given throughout their analysis of the Boyle/Hobbes dispute. What is puzzling in their analysis is the lack of any recognition on their part that just because there exist alternative theoretical explanations for a given collection of observational data doesn't mean that all those alternative explanations are equally plausible and equally likely to be correct. Shapin and Schaffer seem to believe that there can be no principled way of ranking competing alternative accounts of observational data; therefore, they take it as established that *if* there are, say, six alternative and mutually incompatible theoretical explanations of a particular observational datum, there can be no question of one of those six theoretical explanations being any better than the other five. But why should anyone believe that? *Not all theoretical alternatives are equal.* Some alternative explanations cost too much in terms of changes we would have to make in other beliefs. Some alternative explanations postulate new objects or properties that we have no other reason to suppose exist, and so the postulation merely to explain this one observational data-set is *ad hoc.* Some alternative explanations increase the structural complexity of the system under investigation over what other alternative explanations do, and so the former explanations run a higher prior probability of being false. The ways in which competing alternative theories can be unequal with respect to their consequences for methodology, epistemology, and ontology are legion. Shapin and Schaffer seem unaware of this feature of competition between theories. They jump from the fact that Hobbes offered an alternative theoretical account of Boyle's air-pump experiments to the conclusion that Hobbes's account is just as plausible as Boyle's account. But there is no reason to suppose this kind of equality exists among competing theories. In fact, we know that it is extremely difficult to show that two theories are *exactly observationally equivalent.* The only known cases come from rare episodes in the history of physics. But even if two theories can be shown to be exactly observationally equivalent, it doesn't follow from such equivalence alone that the two theories are equivalent with respect to their methodological, epistemological, and ontological implications. We saw [earlier] how Quine, who is very fond of the underdetermination of theory by observational data, nevertheless presents a host of principles that can be used to rank competing theories: simplicity, conservatism, modesty, and fertility. Arguments can be given for why theories that maximize simplicity, consistency with our established beliefs (conservatism), modesty, and fertility at generating novel predictions are more likely to be self-correcting toward the truth over the long run than theories that do not maximize those pragmatic virtues. Hobbes's alternative account of the air-pump experiments did not win out over its rivals in the end. In fact, no theories that postulated a subtle fluid aether—of which Hobbes's was a variety—survived into the twentieth century. They had too many other epistemological costs that would have made it unwise to keep them afloat. Shapin and Schaffer conveniently leave out this subsequent history of atmospheric science. The suggestion at the heart of their study is that if there had been a few key differences in the sociopolitical factors that made Boyle's view triumph, then we would be Hobbesians this very day as you read this—that Hobbes's theories failed not because they were false but because the contingent politics of early science went against them. Scientific truth is a function of human sociopolitical intrigue much more than it is a function of what the objective structure of the world is like, is Shapin and Schaffer's point. As they say near the end of their study, it is "not reality which is responsible for what we know" (*Leviathan and the Air-Pump*, 344).

Once again, the principle of the underdetermination of theory by observational data is being asked to support a form of antirealism that it does not logically entail. This argumentative maneuver is a standard strategy of social constructivists in every discipline. Their assumption seems to be that if nature itself does not unambiguously point to a unique theoretical explanation for something, then there can be no principled way of claiming that there can be an objective truth about that something. But this conditional is unfounded. We can now say with supremely high confidence that:

1. Boyle was correct that the air has weight.
2. Boyle was correct that vacuums are possible.
3. Boyle was incorrect that the air has "spring."
4. Boyle was incorrect that the air contains a subtle aether.
5. Hobbes was incorrect in certain claims about the air (it contains a subtle aether, vacuums are impossible, and all chemical phenomena are mechanical in nature).

Shapin and Schaffer would apparently have us believe that these truths are neither objective nor apolitical; and Latour and

Woolgar would add that the above truths only seem to be objective and independent of sociopolitical determination because the official mythology of science covers up their socially constructed pedigree. It is neither true nor false, per se, that the air has a spring, according to the social constructivists—at best, it is false *relative* to the sociopolitical framework of our current scientific form of life. But, had that sociopolitical framework and its attendant form of life been different in certain ways, then it would be true that the air has a spring. This last claim might strike others besides myself as more than just a little fantastic. But if it were true, then the science critic is potentially greatly empowered to change society through changing its dominant form of knowledge gathering: science. If the facts—all the facts—can be made different by making the social conditions of their construction different, then it is a short step for those social constructivists who harbor an activist streak to see their critiques of science as a form of revolutionary social activism. That is one reason why there is a certain kind of science critic who doesn't mind a model of science in which the input of an independent material world has been reduced to insignificance or eliminated altogether, who is in fact very glad to have arrived at that result. For, to such a science critic, the end product of scientific endeavor under such a model of science is entirely or almost entirely a function of complicated social behavior; and, given that the content (or almost all the content) of science is therefore determined socially, that content is potentially alterable by altering the social conditions of its production. Thus, the possibility arises of altering the substantive content of science through political activism that seeks to alter the social conditions under which the production of scientific knowledge takes place. Some social constructivists, such as Bloor, are not specifically interested in this further activist aspect of the sociology of knowledge. They seem satisfied to point out the social determination processes by which the content of science is produced. We might describe this by saying that those social constructivists stress the *social* in social constructivism. Other social constructivists, in particular [some] feminist science critics stress the *constructivism* aspect of social constructivism, the view that the factual content of science can to a considerable extent be constrained by self-conscious social engineering, by deliberate intervention in the scientific process for purposes of furthering a partisan political agenda. What I want to suggest here is how the nonfeminist social constructivist accounts of science naturally lead, via openly idealist variants of the sociology of knowledge, to the more radically interventionist feminist models of science.

FOUR

A CRITIQUE OF SHAPIN AND SCHAFFER

PAUL R. GROSS AND NORMAN LEVITT

Cultural constructivist theories of science have lately infested the usually staid domain of the history of ideas. One well-known example is the work of Shapin and Schaffer, whose book *Leviathan and the Air-Pump* has a wide circle of admirers. This work is rather more orthodox, on a superficial level, than Latour's. It is an intellectual history of some of the resounding disputes that surrounded the birth of "experimental" science—physics in particular—in the last half of the seventeenth century. What particularly concerns Shapin and Schaffer is the quarrel between some of the most prominent founders of the Royal Society—Boyle, Hooke, and their circle—and the philosopher Thomas Hobbes, author of *Leviathan*. This is the fulcrum upon which they attempt to push the case that, contrary to its flattering image as a uniquely wide-open and tolerant enterprise, welcoming of all new facts, information, and ideas that bear upon its investigations, modern science has been from the first the province of a tightly organized, well-insulated coterie, jealous of its prerogatives and hostile toward outsiders who intrude without the proper credentials. Moreover, this self-appointed scientific aristocracy is seen as organically connected to the ruling elite of Western society. Its views are derived, albeit subtly, from the dominant metaphors of that elite. By the same token, its prestige, authority, and epistemological monopoly are guaranteed by the power of the state and the social formations it principally serves. The argument between Hobbes and the adherents of the Royal Society is offered as an instance of this phenomenon:

> The restored regime [i.e., that of Charles II] concentrated upon means of preventing a relapse into anarchy through the discipline it attempted to exercise over the production and dissemination of knowledge. These political considerations were constituents of the evaluation of rival natural philosophical programmes [i.e., that of the Royal Society's experimentalists, as opposed to the a prioristic rationalism of Hobbes].
>
> Thus the disputes between Boyle and Hobbes became an issue of the security of certain social boundaries and the interests they expressed.[1]

The heart of the matter, as far as Shapin and Schaffer are concerned, is that the confrontation illustrates the degree to which Boyle and his friends were concerned not only with scientific issues, in the narrow sense, but also with the question of credentials. Their supposedly empirical rules, it is said, constituted a specific social practice. They were preoccupied with the question of who should count as a scientific authority, whose judgment was to be respected in scientific disputes, whose evidence was to be accepted as reliable, whose minds were to be acknowledged as sufficiently unpolluted by common prejudice that their observations could be taken at face value.

If we are to believe the Shapin-Schaffer thesis, worthiness to participate in learned discussion of experimental philosophy was closely correlated to rank, wealth, religious orthodoxy, and, in terms of Restoration doctrine, political reliability. This exclusivity was reinforced not only by the money, status, and political connections of many of the members of the Royal Society and their patrons, but in addition by their exclusive possession of the physical instruments of the new experimental method. The air-pump of the title was not a common device. Only a handful existed during the 1660s, and thus the possibility of investigating experimentally the emerging theories of the weight and pressure of gases, now associated with Boyle's name, was limited to the corresponding handful of people who had access to one.

Hobbes, ever the gadfly and eager controversialist, was only too happy to point out this flaw in empiricism. The viciousness of the response to his challenge is to be explained not simply by the theoretical threat it posed to the self-assumed authority of Boyle, Hooke, Oldenburg, and the rest, but, as well, by Hobbes's dark reputation as atheist, philosophical materialist, and general subverter of the sanctity of authority. He was the natural target of distrust because of his lingering reputation as a duplicitous sycophant, willing to flatter either crypto-Catholic king or radical Protestant regicide as the opportunity of the moment suggested. Even more important was his enmity toward religious orthodoxy, and therefore toward the stability of a hierarchical society. The attempt to exile him from the realm of Natural Philosophy therefore must be seen as an act of political prophylaxis.

On this, the Shapin-Schaffer view, the nascent Royal Society was, from the first, the creature and deputy of a political and social viewpoint. The society's supposedly objective science is thus to be read, in large part, as a construction of its ideological commitments, which rejected simultaneously the republican sentiments and leveling enthusiasm of the most radical Puritans and the unconstrained absolutism of the Stuart monarchy. Shapin and Schaffer accept the idea that Hobbes was identified with both kinds of threat.[2] As a defender of absolutism, he could be read as the proponent of a government of unconstrained sovereign power. Yet his fierce independence, which devolved at times into a taste for rancorous disputation, was reminiscent of the intellectual licentiousness of the religious and social radicals of the Civil War period. Given this perspective, the scientific community led by Hooke and Boyle, which echoed the aspirations of a moneyed class that sought immunity from the whims of royalist autocracy while casting a suspicious eye on the tumultuous mass of the unpropertied, had no place for the likes of Thomas Hobbes.

It is not hard to transcribe this view to a contemporary context, as Shapin and Schaffer undoubtedly wish us to do. The analogies are clear. Modern orthodox science is also obsessed by "credentials" in the shape of formal training, academic degrees, and a long period of acclimation to the reigning "paradigms." It polices dissidence and safeguards its monopoly by an elaborate educational system and a forbidding insistence on "peer review." It flourishes with the connivance and support of the organized forces of wealth and authority as constituted in the state, in huge corporations, and in supposedly philanthropic foundations. It has exclusive control over the instruments of empirical investigation, some of which—like multibillion-dollar particle accelerators and orbiting observatories—are far less accessible to the uninitiated than was Boyle's air-pump. And it has its heretics.

Here is Shapin and Schaffer's last word on the general epistemic principle that their particular historical study is supposed to illustrate: "As we come to recognize the conventional and artificial status of our forms of knowing, we put ourselves in a position to realize that it is ourselves and not reality that is responsible for what we know."[3] So, in the end, we come back to the dichotomy—fallacious in that it posits total opposition between "reality" and "convention" where there is, in fact, intense and continuing interaction—so favored by Latour and other constructivists.

The questions raised by *Leviathan and the Air-Pump* are serious and genuine. No intellectually astute history of the interplay between science and its supporting social matrix could afford to ignore them. The flaw, however, lies in attributing a deep and irretrievable source of error to what is ephemeral, local, and inconsistent in its operation. Let us examine the particular picture of seventeenth-century scientific life offered us by Shapin and Schaffer.

Were the panjandrums of the Royal Society really so rigid and intolerant in deciding who was to be accepted as a Natural Philosopher in good standing? Was it true that "the social order implicated in the rationalistic [i.e., a prioristic] production of knowledge threatened that involved in the Royal Society's experimentalism?"[4] It's hard to believe! Recall the roster of thinkers, Continental and English, who were heard with deep respect in the scientific and philosophical debates of the period: In addition to Boyle and Hooke, we have Descartes, a French Catholic; Spinoza,[5] a lapsed Jew doubtful of all religious orthodoxies; and the Royal Society's own Halley, undoubtedly an atheist. Above all, we have Newton himself—by no means a man of property, having little in the way of family connections, and a radical Protestant of fanatical intensity. Newton's singular religious views, be it recalled, prevented him from seeking ordination in the Established Church, a step he diplomatically avoided by becoming Lucasian Professor at Cambridge. Newton was, in fact, an abjurer of the doctrine of the Trinity and thus from many points of view a *heretic*. He was a hostile, secretive, jealous recluse suffering intermittently from mental instability, an unrelenting enemy to the Stuart monarchy in its attempts to sponsor Catholic scholars at Cambridge (thereby opposing its attempts to exercise "discipline . . . over the production and dissemination of knowledge"). To top it all, he was probably homosexual.

Yet consider the celerity with which he was not only embraced but virtually deified by the English intellectual elite, once it became clear that his incomparable mathematical skills had led him to those insights into the nature of physical reality that to this day remain staggering to comprehend. Consider, in particular, the rather touching story of the publication of the *Principia*. It will be remembered that Halley had to drag it out of Newton by main force (imagine a comparable situation involving a contemporary scientist). And Halley, a man of no wealth, put up his own money to see the work through press, taking his compensation in the form of copies, which he had to sell himself. Recall, once more, that Halley was, in fact, an *atheist*, while Newton, on his own testimony, hated atheism above all things! Clearly, there is more to be said about rigidity and latitudinarianism, intolerance and liberty of opinion, in the seventeenth-century scientific community than that the Royal Society constituted a kind of thought police.

Consider again the question of Hobbes's banishment from the circle of the scientific elect. How accurate and complete is Shapin and Schaffer's analysis of the dispute between Hobbes and his foes in the Royal Society? From time to time, they advert to Hobbes's drawn-out fight with the Oxford mathematicians Ward and Wallis, as though its technical aspects were peripheral to their central thesis. They note the existence of the acrimony, and the readiness of the devout Wallis to bring Hobbes's ostensible irreligion into it; but they say nothing about the mathematical substance, claiming that it would carry them too far afield! But of course this is a *central* and highly illuminating question!

Hobbes, be it recalled, had little mathematical training in his youth. He took up the study of Euclidean geometry for the first time in his forties (mathematicians are often said, with some justice, to be washed up at the age of forty) and was an old man at the time of these controversies. Wallis, on the other hand, was, aside from Newton himself, the greatest English mathematician of the seventeenth century. A partisan of Parliament during the Civil War (in anticipation of Alan Turing,[6] he served as code breaker for the Puritan forces), Wallis was an ordained cleric, though of Presbyterian, rather than radical Puritan, leanings. He opposed the execution of Charles I, however, and migrated politically to a position of support for the Restoration settlement. Politics and theology aside, Wallis was a superb, creative mathematician, in contrast to Hobbes, who was, to put it bluntly, incompetent—utterly out of his depth in dealing with subtle mathematical matters.

The controversy between Hobbes and Boyle on the weight and pressure of air must be viewed to a considerable extent as an episode in the twenty-year wrangle between Hobbes and Wallis. It was Wallis who published the most pointed rejoinders to Hobbes, not Boyle himself. The animus between the old philosopher and the Oxford mathematician had arisen from Hobbes's futile criticism of Wallis's mathematics—particularly his great work on infinite series—which antedates the "experimental philosophy" dispute by a number of years. Subsequent to the attack on Boyle's physics, Hobbes once again turned his guns on Wallis's mathematics. But the most revealing, as well as the most comical, quarrel arose when Hobbes published his incorrect solutions to the ancient problems of "squaring the circle" and "duplicating the cube."[7] Wallis, of course, demolished the poor old philosopher's pretensions, and Hobbes compounded the sin, in the eyes of posterity, by being unable (or unwilling) to see the point of Wallis's refutation.

The relevance of these facts to the Shapin-Schaffer hypothesis is that this long and (to Hobbes's admirers) lamentable history provides a concrete and substantive reason, *in contrast to an ideological one*, for Hobbes's notoriety in scientific circles. So far as mathematics is concerned, Hobbes was simply dead wrong in these exchanges, as any competent mathematician would have seen. It is then no wonder that his authority to pass judgment on scientific matters was not well regarded, even if those matters had nothing directly to do with squaring the circle or the like. He was, after all, a strenuous advocate of a rational-deductive methodology *based on that of synthetic geometry*, as an alternative to the emerging experimental empiricism. Shapin and Schaffer emphasize this fact, but unaccountably fail to link it to the question of Hobbes's doubtful mathematical competence.[8] His grotesque failures as a would-be geometer, however, can hardly have been irrelevant.

Leviathan and the Air-Pump would have been a rather different book had it addressed these matters directly. The image of Hobbes as brilliant and devastating iconoclast would have taken some hits, at the least. Moreover, Shapin and Schaffer would have put themselves in the position of conceding the existence of sound, objective reasons for deciding at least some scientific controversies—that between Hobbes and Wallis being an important case in point. Inevitably, they would have been led to concede that there are reasonably valid criteria for deciding the scientific competence of individuals, for distinguishing, in most instances, between worthwhile theorists and cranks. After all, in terms of mathematics, Hobbes was a crank. Such concessions, however, do not sort well with a relativist or conventionalist position, especially one grounded on a radically antielitist politics. Shapin and Schaffer sidestep the issues that might entail such admissions by insisting that all such disputes are ideological.[9]

Leviathan and the Air-Pump is exhaustively and meticulously researched as a narrative of events and personalities during a short span of time. Nonetheless, the ideological perspectives of its authors make it an exercise in tunnel vision. To concentrate on the idea of empirical science as a manifestation of cultural and political imperatives is to omit important dimensions of the story, both human and philosophical. The efforts of Boyle and his colleagues to put science on a solid experimental footing and to restrain the impulse toward a priori speculative systems was a project facing substantial practical difficulties at that stage. It is one thing to embrace "empiricism" in the abstract, quite another to find practical and reliable methods for developing and extending concrete knowledge. The early experimental philosophers were confronted with the necessity to minimize the effects of human fallibility and bias, and it is shortsighted to condemn them out of hand for addressing the difficulties in language that occasionally smacks of snobbery or political insecurity. The verdict of history must be that they succeeded magnificently in sketching the broad methodological outline by which the physical and biological sciences have attained their present scope and power. To put it another way, irrespective of the "social" grounding of their ideas, what more could they possibly have done, short of inventing the theory of experimental design and developing the techniques of mathematical statistics and error theory that underlie it?

Furthermore, it is false to read their rejection of Hobbes as a blanket denial of the value of speculative and deductive thought. Such reasoning was eagerly received when it was the product of genuine intellectual competence, as in the case of Huygens and, of course, of Newton himself. The singular genius of the period was to explore new and powerful mathematical reasoning in the service of physical science *without* falling into the trap of contempt for mere experience. The authors concede that they have not quite come to understand how the experimentalism of Boyle was made to dovetail with the mathematical science of Newton; but this may well be because they have been celebrating the wrong hero. Hobbes, the mathematical dilettante and bumbler, simply does not belong in the same pantheon with Descartes, Huygens, Newton, Leibniz, and Bernoulli. His misadventures are tiresome and, in the last analysis, uninstructive.

A final word about the rhetoric of the book: Once more, we find an argument designed to appeal to a certain kind of readership on grounds other than strict logic and evidence. To side with Boyle and the Royal Society crowd, as the book presents them, is to side with snobbish, purse-proud, rank-conscious plutocrats in their fear of the disorderly masses. If Hobbes cannot be construed as radical democrat (as indeed he cannot—his motivations are markedly authoritarian), then at least he can be made to stand for the voiceless and excluded masses—and the intellectuals without serious scientific training—to whom science is an inaccessible mystery, seemingly beyond human control. Thus we are forced in our reading of the book to see it as a parable, whose fulsome celebration of Hobbes conveys the implication that "philosophers" who are not professional scientists (for which we must read "historians" and "sociologists") should have the authority to pronounce, or even to prescribe, on scientific questions. As we have observed before, this kind of stacking of the emotional deck has great persuasive force on the academic left, irrespective of the soundness of the argument that encodes it.

NOTES

1. Steven Shapin and Simon Schaffer, *Leviathan and the Air-Pump, p.* 283
2. Ibid., p. 301.
3. Ibid., p. 344.
4. Ibid., p. 139.
5. Shapin and Schaffer identify Spinoza along with Hobbes as an opponent of Boyle's particular views of the nature of air, as well as a builder of a priori rationlistic systems, and thus an opponent of "experimental" philosophy. Spinoza clearly was an a priorist—and much better than Hobbes as a pure dialectician in this vein. On the other hand, as a leading expert on the science of optics—and a professional lensgrinder—Spinoza must also count as an "experimentalist." Still, given Spinoza's "outcast" status, as well as his subversive views on religion, he should have been cast further into the outer darkness even than poor old Hobbes! Yet he was closely associated with Henry Oldenburg, the Royal Society's inaugural secretary.
6. See Andrew Hodges, *Alan Turing: The Enigma* (New York: Simon and Schuster, 1983).
7. These problems—constructing a square equal in area to a given circle or a cube of volume twice that of a given cube using only the ideal compass and straightedge of synthetic Euclidean geometry—go back to classical antiquity. By means of deep algebraic arguments, they were shown, in the nineteenth century, to be insoluble.
8. Shapin and Schaffer, *Leviathan and the Air-Pump*, p. 135. Also cited (p. 100) is Hobbes's question, "For who is so stupid, as both to mistake in geometry, and also to persist in it, when another detects his error to him?" but not the ironic answer, "Thomas Hobbes!"
9. In taking this position, we contradict not only Shapin and Schaffer, but some later work on the same subject from a similar social-determinist perspective. "Why was Hobbes excluded from membership in the Royal Society?" asks James R. Jacob in "The Political Economy of Science in Seventeenth-Century England," *Social Research* 59, no. 3 (1992): 532n. "Quite simply," he answers, "we can now see that the social and political views of leaders of the Society like Boyle and Wilkins diverged sharply from those of Hobbes in certain fundamental respects." Again, out of eagerness to derive everything from "politics," a scholar has overlooked the contrast between Hobbes's high opinion of his own mathematical abilities and his manifest mathematical incompetence. Hobbes's contemporaries, however, were not deceived on this point.

STUDY QUESTIONS

Latour and Woolgar speak of "inversions" that occur when a scientific hypothesis is no longer considered merely hypothetical, but is viewed as an accurate depiction of reality. Such "inversions" occur not just in science but in everyday life. For instance, when someone is accused of a crime, there is (or should be) a presumption of innocence, and the suspect is referred to as the "accused" or "alleged" perpetrator. Such language is used to remind us that guilt is at that point only hypothetical and not yet established. After conviction, however, the guilty party is no longer merely accused but is regarded as an actual criminal. What really goes on when such an "inversion" occurs? Do we all, as Latour and Woolgar suggest, suffer a collective amnesia and forget the rhetorical means whereby our "facts" are "constructed?" Consider an "inversion" in everyday life and trace how it occurs.

Shapin and Schaffer argue, in effect, that "the winners write the textbooks." That is, when there is a dispute about scientific methods and procedures, the winning party gets to impose its position on the scientific community and ensure that the next generation of scientists will be taught that the winners' view is the *right* one. Obviously, there is *some* truth in this claim, but just how much? Is it really conceivable that science since the time of Hobbes and Boyle could have been done without laboratories or experiments?

Klee argues that Latour and Woolgar make a basic mistake. He accuses them of thinking that because science often employs elaborate and complex devices, anything scientists claim to detect with such devices is actually an artifact of the device itself. Now scientists themselves have long recognized that sometimes it is hard to distinguish the "signal" from the "noise." That is, it is often hard to distinguish real phenomena from artifacts of the detecting devices. Yet scientists believe that they have many checks and safeguards that permit them to do so. Are Latour and Woolgar best interpreted as claiming that such safeguards are always inadequate, or are they perhaps claiming something deeper? Could one defend the view that, in some sense, scientific objects are created by scientific inquiry? Would any such claim imply that, for instance, the Galilean moons of Jupiter did not exist until Galileo detected them?

Gross and Levitt hold that Hobbes, in his dispute with Boyle and the mathematician Wallis, was little more than a crackpot. His rejection by the scientific establishment was therefore justified and not a matter of prejudice. However, scientists often argue among themselves about the proper methodologies and techniques. How, other than through political wrangling, can debates about method be settled when they arise from within science itself?

PART 2

FEMINISTS ON SCIENCE

INTRODUCTION

When you flip through the index of any standard history of science and look at the names, you recognize a considerable diversity. Over the last 2,500 years, significant scientific discoveries have been made by Chinese, Indians, Arabs, Persians, Mayans, Italians, Greeks, French, Germans, English, Scots, Americans, Russians, Danes, Poles, Hungarians . . . the list could go on and on. Science has been and is being done by Muslims, Hindus, Taoists, Buddhists, Christians, Jews, atheists, agnostics, and pagans. Scientists have included persons of almost every conceivable race or ethnicity. Surely science is open for all to pursue; it is quintessentially *public* knowledge, not the exclusive domain of any particular clique or cadre. Yet for all the diversity of scientists, one glaring fact stands out: The overwhelming majority of scientists have been, and are, male. Of course, one can produce quite a respectable list of famous female scientists: Hypatia, Caroline Herschel, Marie Curie, Irene Joliot-Curie, Lise Meitner, Barbara McLintock, Cecilia Payne-Gaposchkin, Rosalind Franklin, and Lynn Margulis come immediately to mind. Still, in the various fields of the sciences, and especially engineering, men have been numerically dominant.

Why is this so? Do men possess "science and math" genes that women lack? It hardly needs to be said that no such masculine genetic advantage has been found. Why then, starting around the fourth grade, do girls start to fall behind boys in science and math achievement? Overall, girls get better grades than boys, yet by ninth grade girls show significantly less interest in and aptitude for science and math. Are girls more attracted to the "more human" fields of art and literature and repelled by the perceived austerity or coldness of science? *If* this is so, it merely raises the deeper question of why it is so.

Feminists have answers to these questions: Science is not only run by but for males; it is a boys' club. Women have been actively discouraged from pursuing scientific or engineering careers. Several women of my personal acquaintance were told, to their faces, that "women can't do chemistry," or "women don't make good engineers." A supervisor told one woman, a nuclear engineer, that women are "weak links" who must be driven out of the profession. Presumably, the makers of these statements felt that women should be home baking cookies and not worrying their pretty little heads with differential equations. Other forms of discrimination are less blatant. Female scientists and engineers can simply be "left out of the loop" or assigned peripheral and lower-status jobs.

Feminists also argue that there are deeper reasons that many women are alienated from science. They charge that scientific research is often conducted under biased assumptions that slant scientific results against women. There is no doubt that prejudices have sometimes skewed science, especially in fields that bear on our self image. Stephen Jay Gould's book *The Mismeasure of Man* amusingly and horrifyingly shows how European scientists, pursuing an allegedly objective methodology, systematically distorted data to "prove" that Asians and Africans had less brain capacity, and so supposedly less intelligence, than white Europeans. Some German scientists even found that Germans generally had larger brains than other Europeans! These conclusions were later shown to be rubbish, but why were they accepted in the first place? The answer is obvious: Scientists are subject to prejudices of all sorts like other people, and not always good at devising methods to eliminate that bias from their research.

Consider primatology, the field that studies apes and monkeys. Primates are generally highly social creatures that often live in small groups of no more than a few dozen individuals. Such groups usually have an "alpha male," a male that physically dominates the other males and has the primary (or exclusive) right to mate with females. The "silverback" gorilla is an example of such an alpha male. Early work by (mostly male) primatologists emphasized the role and importance of alpha males in protecting and leading their groups. Other group members, both male and female, were depicted as meekly submissive to the alpha male's authority. However, later research showed that female primates often have power that limits the alpha male's rule. For instance, among mandrills, when a new male takes over the top role, the females are not instantly his to command. Though he is much larger and stronger than any individual female, and has beaten all male rivals, the females sometimes defiantly refuse to mate with him.

Why did the (mostly male) primatologists first overstate the dominance of the alpha males? For feminists the answer is obvious. Because primates resemble us so, and because evolutionary theory locates them on adjoining branches of the tree of life, it is very hard not to read the rules for human society into their social behavior. In other words, the male dominance found in human societies is projected onto primate groups. There is then a process of reverse reasoning in which, because male domination is seen as the rule among our closest animal relations, it is taken to be the "natural" state for humans. By such a circular process, feminists charge, the human cultural practice of male domination seeks "scientific" validation.

Such charges of sexism may be inflammatory, but they do not undermine the basic objectivity of the scientific process, even if they note its failure in specific cases. On the contrary, if sexism or other forms of prejudice have infected certain branches of science, exposing and eradicating such prejudice can only help those fields advance. Of course, the charge that sexism has skewed theories in some fields, like primatology, anthropology, or psychology, is more plausible than the same charge applied to, say, particle physics. Particle physics may have been, and perhaps to some extent still is, a "boys' club," but it is hard to imagine that female physicists would have found different particles obeying different laws of nature. Would women have found a different charge for the electron or another mass for the proton? Would Maxwell's equations or Avogadro's number or Newton's laws of motion have been substantially different had women formulated them rather than those particular men? Are the laws of thermodynamics sexist? It just is not plausible to see the influence of sexism as extending that far. Yet some feminists have taken their critiques that far and farther.

Some feminist theorists have made extreme claims, arguing that science, in its basic methods, standards, and values, is shot through with patriarchal and oppressive assumptions. They see science as a product of Western, linear, male (all bad things) ways of thinking, and logic itself as a tool of coercion. Even the scientific ideals of objectivity and impartiality have been derided as smokescreens hiding an ideology of male dominance. The young feminist quoted in the introduction who said, "Objectivity is what a man calls his subjectivity," was, I think, trying to assert such a charge. At its outer fringes, feminism merges with New Age themes of goddess worship and toys with very dubious notions, such as that women have an innate, mystical bond with nature that men lack. Other feminists have criticized these ideas as harking back to sexist stereotypes about women's allegedly more "intuitive" or "spiritual" character.

Since feminists raise many touchy questions, brashly politicizing concerns that had previously been regarded as strictly personal, a penumbra of raw emotion surrounds feminist issues. Extremist rhetoric has been deployed both for and against feminism. One right-wing radio commentator derides feminists as "feminazis"; a well-known religious broadcaster defined feminism as the view that women should abandon their husbands and children and become witches and lesbians. Unfortunately, the behavior of some feminists and certain of the more extreme claims made in the name of feminism have encouraged such hostile caricatures. One well-known feminist professor and lecturer refuses to admit men into her classes or to acknowledge questions from male audience members. Heterosexual feminists have been chastised by more radical colleagues for "sleeping with the enemy." One prominent feminist legal scholar would erase the distinction between pornographic speech and the act of rape.

It is important not to let extreme rhetoric or far-out claims lead us into hostile stereotyping. Feminism is no more a monolithic ideology than, say, conservatism. "Conservative" and "feminist" are vague terms encompassing a wide variety of positions. What unites feminists (or conservatives) is a loose set of interrelated beliefs, perspectives, and values. Minimally, I would say, feminists believe that women have historically been unjustly excluded from positions of power, prestige, and leadership, and denied opportunities of personal enrichment, achievement, or self-fulfillment usually available to men. Feminists further hold that current social policy should actively seek to right and redress those wrongs. So defined, feminism is clearly a big tent, and many different opinions on fundamental issues are to be expected. True, some feminists have delved deeply into relativism, postmodernism, and even New Age mysticism. Other feminists have vigorously opposed these trends. For instance, Noretta Koertge and Daphne Patai, formerly active in the movement to promote women's studies, wrote the book *Professing Feminism* (1994), protesting the direction such studies were going. So, the readings for this section should be seen not as feminists versus antifeminists but as a debate or dialogue among feminists.

The readings for this section feature a lengthy selection from Sandra Harding, perhaps the best-known feminist philosopher of science. The reading is taken from her book *Whose Science? Whose Knowledge? Thinking from Women's Lives.* Harding first defines and defends "feminist standpoint epistemology." Epistemology is the branch of philosophy concerned with the nature, scope, and basis of knowledge. Epistemologists are especially interested in how some, but not all, of our beliefs qualify as knowledge. To count as knowledge, a belief must be true, but that is not enough. True beliefs must also be justified or warranted, that is, there must be adequate rational grounds for holding them to be true. Epistemologists seek to specify the necessary and sufficient conditions for beliefs to acquire such justification or warrant.

Epistemologists have traditionally sought objective, universal criteria that demarcate the rational from the irrational, and distinguish well-grounded or strongly confirmed claims from those that are merely speculative or conjectural. Because they seek universal and objective criteria, traditional epistemologists think that the particular perspectives or standpoints of groups or individuals are irrelevant. Such parochial or personal viewpoints must be transcended by appeals to pure reason, which is held to be the same for all rational beings regardless of class, gender, ethnicity, and so on. It follows on the traditional view that the mind has no gender, that is, the dictates of reason are universal and unaffected by such contingencies as the gender of the knowing subject.

Feminist standpoint epistemology strongly disagrees with this traditional view. It contends that women's experience, articulated from a specifically *feminist* perspective, has *cognitive* significance. That is, the experience of women as the victims of oppression, once brought into self-consciousness by feminist analysis and reflection, can give women privileged access to truths denied to men. Just as the slave knows all of the master's quirks, moods, and foibles, while the master is largely ignorant of his slaves' lives, so women, as victims of gender discrimination, know much that men cannot see.

Harding realizes that this claim will strike many as harking back to old stereotypes about women's allegedly superior intuition. She also knows that she is going against over two thousand years of philosophical tradition. She therefore offers eight reasons (the first four are presented in the following selection) why feminists can be expected to have superior insight. Her presentation of these reasons is clear, so there is no need to summarize them here. In general, these arguments emphasize that women's lives have been undervalued and their experiences dismissed by the dominant (largely male-constructed) intellectual culture. She contends that empirical knowledge of the world, which, by definition, is based on human experience of the world, is stunted if the characteristic experiences of more than half of humanity are excluded from our data.

If Harding's argument has succeeded so far, it is still not clear what is implied for *epistemology*. As noted above, epistemology is concerned with the criteria or standards of justified belief. Even if women's experience, as interpreted by feminist analysis, does give some individuals superior insight, it is not clear how this affects our views about *what counts* as knowledge or our *standards* of justification. Harding answers that feminist standpoint epistemology offers science a *stronger* form of objectivity. Traditional epistemologists hold that beliefs can be objective only if the grounds for holding them are scrupulously guarded against the polluting influence of politics or ideology. Harding holds that, on the contrary, epistemic criteria grounded in the feminist standpoint will be *more* objective! Harding, unlike some postmodernists, is not a science basher. She claims to want to make science better able to depict nature as it is.

Harding calls "weak objectivity" the kind of objectivity that has so far been the scientific ideal. She characterizes such "weak objectivity" as the ideal that science should be value-free, disinterested, and neutral with respect to all ideologies or political agendas. Her argument against "weak objectivity" is that science never has and never will be objective in that sense. In fact, she charges, whenever science has claimed such objectivity, this claim has only masked the real agendas, ideologies, and interests deeply embedded in scientific practice. She holds that Thomas Kuhn and others have shown that science is inevitably pervaded by all sorts of political, cultural, philosophical, and ideological priorities. The solution is not to pretend that these "extraneous" influences can be excised by doing "better" science, but by consciously doing science from assumptions and perspectives that empower the marginalized. Instead of doing science in a way that promotes the interests of a white, male, bourgeois elite—as it has usually been done—science should consciously adopt perspectives that liberate the oppressed and reach out to the excluded. Only such a perspective will give science "strong objectivity" rather than the spurious "weak objectivity" it has idolized.

The next two readings are criticisms directed at Harding and feminist epistemology. The first is Cassandra Pinnick's article "Feminist Epistemology: Implications for Philosophy of Science." Pinnick argues that Harding's attack on traditional ("weak") objectivity assumes that historians of science such as Thomas Kuhn and philosophers such as W. V. O. Quine have shown that science is a nonrational enterprise governed by politics, ideology, and the like. She notes that many of the most prominent philosophers of science have argued in detail that this has not been shown. Pinnick also considers the claims of the Strong Program in the sociology of knowledge. Advocates of the Strong Program hold that sociology can explain even the most successful episodes in science, like the discovery of the double-helix structure of DNA. That is, the results of science are not seen as due to nature but explained as the product of the conventions and practices of scientific communities, which themselves were adopted to serve social, cultural, and political imperatives. Scientific "discoveries" therefore are not facts but artifacts. Harding invokes the claims of the Strong Program to support her criticisms of "weak objectivity." Pinnick argues that the Strong Program fails to establish that social, rather than traditionally objective, factors caused the success of the winning theories in science. Therefore, the Strong Program provides no support for Harding's critique.

Pinnick also considers Harding's claim that feminists, because of their marginalized status, can have privileged insights into the natural world. Pinnick regards this as an empirical claim—one to be settled by evidence. She says that Harding simply has not provided such evidence. Harding offers no data to show that feminists have made significant contributions to any major field of science. Pinnick might have added that Harding must also show that successful feminist scientists were successful *because of their feminism,* and not simply because they were excellent scientists who happened to be feminists.

In her book *Feminism Under Fire,* from which the next selection is taken, Ellen Klein also focuses on Harding's views on objectivity. Klein charges that Harding has waffled on her definitions of "strong" objectivity. If Harding means that science is to be based on an explicitly political agenda—even one that emphasizes the primacy of "women's lives"—Klein says that such a stance will, at best, be irrelevant to scientific problems. At worst, it could distort science just as badly as any other ideological bias. If Harding waters down her definition of "strong" objectivity, then it becomes hard to see how her notion differs from traditional concepts of objectivity.

Klein also considers the "weak" objectivity that Harding attacks. If "weak" objectivity implies that scientists must have a neutral or detached attitude toward their scientific work, then Klein agrees that such objectivity is a myth. Scientists are not aloof, passive observers but are avidly involved in the discovery process. They care passionately about their research and often have strong expectations about what they will find. Yet, Klein notes, there is a clear distinction between being *un*interested and being *dis*interested. Objectivity requires the latter, but not the former. Objectivity comes from recognizing that we *are* biased and seeking methods, techniques, and standards of inquiry that permit us to transcend that bias. Klein argues that Harding has *not* shown that objectivity in this latter sense is impossible.

FIVE

FEMINIST STANDPOINT EPISTEMOLOGY AND STRONG OBJECTIVITY

SANDRA HARDING

[Feminist standpoint theories] argue that not just opinions but also a culture's best beliefs—what it calls knowledge—are socially situated. The distinctive features of women's situation in a gender-stratified society are being used as resources in the new feminist research. It is these distinctive resources, which are not used by conventional researchers, that enable feminism to produce empirically more accurate descriptions and theoretically richer explanations than does conventional research. Thus, the standpoint theorists offer an explanation different from that of feminist empiricists of how research directed by social values and political agendas can nevertheless produce empirically and theoretically preferable results. . . .

This justificatory approach originates in Hegel's insight into the relationship between the master and the slave and the development of Hegel's perceptions into the "proletarian standpoint" by Marx, Engels, and Georg Lukács. The assertion is that human activity, or "material life," not only structures but sets limits on human understanding: What we do shapes and constrains what we can know. As Hartsock argues, if human activity is structured in fundamentally opposing ways for two different groups (such as men and women), "one can expect that the vision of each will represent an inversion of the other, and in systems of domination the vision available to the rulers will be both partial and perverse."[1]

The feminist standpoint theories focus on gender differences, on differences between women's and men's situations which give a scientific advantage to those who can make use of the differences. But what are these differences? On what grounds should we believe that conventional research captures only "the vision available to the rulers"? Even if one is willing to admit that any particular collection of research results provides only a partial vision of nature and social relations, isn't it going too far to say that it is also perverse or distorted?[2] What is it about the social situation of conventional researchers that is thought to make their vision partial and distorted? Why is the standpoint of women—or of feminism—less partial and distorted than the picture of nature and social relations that emerges from conventional research?

We can identify many differences in the situations of men and women that have been claimed to provide valuable resources for feminist research. These can be thought of as the "grounds" for the feminist claims.[3]

(1) Women's different lives have been erroneously devalued and neglected as starting points for scientific research and as the

generators of evidence for or against knowledge claims. Knowledge of the empirical world is supposed to be grounded in that world (in complex ways). Human lives are part of the empirical world that scientists study. But human lives are not homogeneous in any gender-stratified society. Women and men are assigned different kinds of activities in such societies; consequently, they lead lives that have significantly different contours and patterns. Using women's lives as grounds to criticize the dominant knowledge claims, which have been based primarily in the lives of men in the dominant races, classes, and cultures, can decrease the partialities and distortions in the picture of nature and social life provided by the natural and social sciences.[4]

Sometimes this argument is put in terms of personality structures. Jane Flax and other writers who draw on object relations theory point to the less defensive structure of femininity than of masculinity. Different infantile experiences, reinforced throughout life, lead men to perceive their masculinity as a fragile phenomenon that they must continually struggle to defend and maintain. In contrast, women perceive femininity as a much sturdier part of the "self." Stereotypically, "real women" appear as if provided by nature; "real men" appear as a fragile social construct. Of course, "typical" feminine and masculine personality structures are different in different classes, races, and cultures. But insofar as they are different from each other, it deteriorates objectivity to devalue or ignore what can be learned by starting research from the perspective provided by women's personality structures.[5]

Sometimes this argument is put in terms of the different modes of reasoning that are developed to deal with distinctive kinds of human activity. Sara Ruddick draws our attention to the "maternal thinking" that is characteristic of people (male or female) who have primary responsibility for the care of small children. Carol Gilligan identifies those forms of moral reasoning typically found in women's thought but not found in the dominant Western "rights orientation" of ethics. And Mary Belenky and her colleagues argue that women's ways of knowing exhibit more generally the concern for context that Gilligan sees in moral knowing.[6]

One could argue also that the particular forms of any emotion that women experience as an oppressed, exploited, and dominated gender have a distinctive content that is missing from all those parallel forms in their brothers' emotional life. Consider suffering, for example. A woman suffers not only as a parent of a dying child, as a child of sick parents, as a poor person, or as a victim of racism. Women suffer in ways peculiar to *mothers* of dying children, to *daughters* of sick parents, to poor *women,* and in the special ways that racist policies and practices affect *women's* lives. Mother, daughter, poor woman, and racially oppressed woman are "nodes" of historically specific social practices and social meanings that mediate when and how suffering occurs for such socially constructed persons. Women's pleasures, angers, and other emotions, too, are in part distinctive to their social activities and identities as historically determinate women, and these provide a missing portion of the human lives that human knowledge is supposed to be both grounded in and about.

Whatever the kind of difference identified, the point of these arguments is that women's "difference" is only difference, not a sign of inferiority. The goal of maximizing the objectivity of research should require overcoming excessive reliance on distinctively masculine lives and making use also of women's lives as origins for scientific problematics, sources of scientific evidence, and checks against the validity of knowledge claims.

Some thinkers have assumed that standpoint theories and other kinds of justifications of feminist knowledge claims must be grounded in women's *experiences.* The terms "women's standpoint" and "women's perspective" are often used interchangeably, and "women's perspective" suggests the actual perspective of actual women—what they can in fact see. But it cannot be that women's experiences in themselves or the things women say provide reliable grounds for knowledge claims about nature and social relations. After all, experience itself is shaped by social relations: for example, women have had to *learn* to define as rape those sexual assaults that occur within marriage. Women had experienced these assaults not as something that could be called rape but only as part of the range of heterosexual sex that wives should expect.

Moreover, women (feminists included) say all kinds of things—misogynist remarks and illogical arguments; misleading statements about an only partially understood situation; racist, class-biased, and heterosexist claims—that are scientifically inadequate. (Women, and feminists, are not worse in this respect than anyone else; we, too, are humans.) Furthermore, there are many feminisms, and these can be understood to have started their analyses from the lives of different historical groups of women: liberal feminism from the lives of women in eighteenth- and nineteenth-century European and American educated classes; Marxist feminism from the lives of working-class women in nineteenth- and twentieth-century industrializing societies; Third World feminism from late-twentieth-century Third World women's lives. Moreover, we all change our minds about all kinds of issues. So while both "women's experiences" and "what women say" certainly are good places to begin generating research projects in biology and social science, they would not seem to be reliable grounds for deciding just which claims to knowledge are preferable.

For a position to count as a standpoint, rather than as a claim—equally valuable but for different reasons—for the importance of listening to women tell us about their lives and experiences, we must insist on an objective location—women's lives—as the place from which feminist research should begin. We would not know to value that location so highly if women had not insisted on the importance of their experiences and voices. (Each woman can say, "I would not know to value my own experience and voice or those of other women if women had not so insisted on the value of women's experiences and voices.") But it is not the experiences or the speech that provides the grounds for feminist claims; it is rather the subsequently articulated observations of and theory about the rest of nature and social relations—observations and theory that start out from, that look at the world from the perspective of, women's lives. And who is to do this "starting out"? With this question it becomes clear that knowledge-seeking requires democratic, participatory politics. Otherwise, only the gender, race, sexuality, and class elites who now predominate in institutions of knowledge-seeking will have the chance to decide how to start asking their research questions, and we are entitled to suspicion about the historic location from which those questions will in fact be asked. It is important both to value women's experiences and speech and also to be able to specify carefully their exact role in the production of feminist knowledges.

(2) Women are valuable "strangers" to the social order. Another basis claimed for feminist research by standpoint thinkers is women's exclusion from the design and direction of both the social order and the production of knowledge. This claim is supported by the sociological and anthropological notion of the stranger or outsider. Sociologist Patricia Hill Collins summarizes the advantages of outsider status as identified by sociological theorists. The stranger brings to her research just the combination of nearness and remoteness, concern and indifference, that are central to maximizing objectivity. Moreover, the "natives" tend to tell a stranger some kinds of things they would never tell each other; further, the stranger can see patterns of belief or behavior that are hard for those immersed in the culture to detect.[7] Women are just such outsiders to the dominant institutions in our society, including the natural and social sciences. Men in the dominant groups are the "natives" whose life patterns and ways of thinking fit all too closely the dominant institutions and conceptual schemes.

In the positivist tendencies in the philosophy of the social sciences, these differences between the stranger and the natives are said to measure their relative abilities to provide causal explanations of the natives' beliefs and behaviors. Only understanding, not explanation, can result from the natives' own accounts of their beliefs and behaviors, or from the accounts of anthropologists or sociologists who "go native" and identify too closely with the natives. Because women are treated as strangers, as aliens—some more so than others—by the dominant social institutions and conceptual schemes, their exclusion alone provides an edge, an advantage, for the generation of causal explanations of our social order from the perspective of their lives. Additionally, however, feminism teaches women (and men) how to see the social order from the perspective of an outsider. Women have been told to adjust to the expectations of them provided by the dominant institutions and conceptual schemes. Feminism teaches women (and men) to see male supremacy and the dominant forms of gender expectations and social relations as the bizarre beliefs and practices of a social order that is "other" to us. *It* is "crazy"; we are not.

This claim about the grounds for feminist research also captures the observation of so many sociologists and psychologists that the social order is dysfunctional for women. There is a closer fit for men in the dominant groups between their life needs and desires and the arrangement of the social order than there is for any women. But this kind of claim has to be carefully stated to reflect the extremely dysfunctional character of the U.S. social order for men who are *not* members of dominant groups—for example, African Americans and Hispanics. It is clearly more dysfunctional for unemployed African American and Hispanic men than it is for economically privileged white women. Nevertheless, with extremely important exceptions, this insight illuminates the comparison of the situation of women and men in many of the same classes, races, and cultures. It also captures the observation that within the same culture there is in general a greater gap for women than for men between what they say or how they behave, on the one hand, and what they think, on the other hand. Women feel obliged to speak and act in ways that inaccurately reflect what they would say and do if they did not so constantly meet with negative cultural sanctions. The socially induced need for women always to consider "what men (or 'others') will think" leads to a larger gap between their observable behavior and speech and their thoughts and judgments.

(3) Women's oppression gives them fewer interests in ignorance. The claim has been made that women's oppression, exploitation, and domination are grounds for transvaluing women's differences because members of oppressed groups have fewer interests in ignorance about the social order and fewer rea-

sons to invest in maintaining or justifying the status quo than do dominant groups. They have less to lose by distancing themselves from the social order; thus, the perspective from their lives can more easily generate fresh and critical analyses. (Women have less to lose, but not nothing to lose; gaining a feminist consciousness is a painful process for many women.)

This argument can be put in terms of what women, and especially feminist women, can come to be willing to say. But it is less confusing if it is put in terms of what can be seen if we start thinking and researching from the perspective of the lives of oppressed people. The understanding that they are oppressed, exploited, and dominated—not just made miserable by inevitable natural or social causes—reveals aspects of the social order that are difficult to see from the perspective of their oppressors' lives. For example, the perception that women believe they are firmly saying no to certain sexual situations in which men consistently perceive them to have said—yes or "asked for it" (rape, battering) becomes explainable if one believes that there can never be objectively consensual relations between members of oppressor and oppressed groups. It is from the perspective of women's interests that certain situations can be seen as rape or battering which from the perspective of the interests of men and the dominant institutions were claimed to be simply normal and desirable social relations between the sexes.

(4) Women's perspective is from the other side of the "battle of the sexes" that women and men engage in on a daily basis. "The winner tells the tale," as historians point out, and so trying to construct the story from the perspective of the lives of those who resist oppression generates less partial and distorted accounts of nature and social relations.

Far from being inert "tablets"—blank or not—human knowers are active agents in their learning. Knowledge emerges for the oppressed through the struggles they wage against their oppressors. It is because women have struggled against male supremacy that research starting from their lives can be made to yield up clearer and more nearly complete visions of social reality than are available only from the perspective of men's side of these struggles. "His resistance is the measure of your oppression" said the early 1970s slogan that attempted to explain why it was that men resisted so strenuously the housework, child care, and other "women's work" that they insisted was so easy and required so few talents and so little knowledge.

As I put the point earlier, knowledge is produced through "craft" procedures, much as a sculptor comes to understand the real nature of the block of marble only as she begins to work on it. The strengths and weaknesses of the marble—its unsuspected cracks or surprising interior quality—are not visible until the sculptor tries to give it a shape she has in mind. Similarly, we can come to understand hidden aspects of social relations between the genders and the institutions that support these relations only through struggles to change them. Consider an example from the history of science: it is only because of the fierce struggles waged in the nineteenth and early twentieth centuries to gain formal equality for women in the world of science that we can come to understand that formal equality is not enough. As Margaret Rossiter points out, all the formal barriers to women's equity in education, credentialing, lab appointments, research grants, and teaching positions have been eliminated, yet there are still relatively few women to be found as directors and designers of research enterprises in the natural sciences.[8] The struggles to end discrimination against women in the sciences enabled people to see that formal discrimination was only the front line of defense against women's equity in scientific fields.

Hence, feminist politics is not just a tolerable companion of feminist research but a necessary condition for generating less partial and perverse descriptions and explanations. In a socially stratified society the objectivity of the results of research is increased by political activism by and on behalf of oppressed, exploited, and dominated groups. Only through such struggles can we begin to see beneath the appearances created by an unjust social order to the reality of how this social order is in fact constructed and maintained. This need for struggle emphasizes the fact that a feminist standpoint is not something that anyone can have simply by claiming it. It is an achievement. A standpoint differs in this respect from a perspective, which anyone can have simply by "opening one's eyes." Of course, not all men take the "men's position" in these struggles; there have always been men who joined women in working to improve women's conditions, just as there have always been women who—whatever their struggles with men in their private lives—have not thought it in their interest to join the collective and institutional struggles against male supremacy. Some men have been feminists, and some women have not. . . .

Standpoint epistemologies are most convincing to thinkers who are used to investigating the relationship between patterns of thought and the historical conditions that make such patterns reasonable. Consequently, many historians, political theorists, and sociologists of knowledge can find these explanations of why feminist research can generate improved research results more plausible than feminist empiricism.

The diversity of the resources that in other forms are familiar in the social sciences, and that feminists can call on in defending the greater objectivity attainable by starting research from women's lives, is another great advantage. It is hard to imagine how to defeat this entire collection of arguments—and the others to be found in feminist research—since they are grounded in a variety of relatively conventional understandings in the social sciences.

Moreover, the standpoint theories, like feminist empiricism, can claim historical precedents. Many (though not necessarily all) of the grounds identified above are used by the new histories of science to explain the emergence of modern science.[9] Scientific method itself was created by a "new kind of person" in the early modern era. Feudalism's economic order separated hand and head labor so severely that neither serfs nor aristocrats could get the necessary combination of a trained intellect and willingness to get one's hands dirty that are necessary for experimental method. One can also point to pre-Newtonian science's involvement in political struggles against the aristocracy. Or one can focus on the "fit" of Ptolemaic astronomy's conceptual scheme with the hierarchical social structure of the Catholic Church and feudal society while, in contrast, the Copernican astronomy mirrored the more democratic social order that was emerging. Or one can note the way the problematics of the new physics were "for" the rise of the new merchant classes: it was not that Newton set out to "conspire" with these classes; rather, his new physics solved problems that had to be solved if transportation, mining, and warfare were to be more efficient.[10] So the feminist empiricists' appeals to historical precedent can be made in a different way by the standpoint theorists. . . .

I [have] argued that a feminist standpoint theory can direct the production of less partial and less distorted beliefs. This kind of scientific process will not merely acknowledge the social-situatedness—the historicity—of the very best beliefs any culture has arrived at or could in principle "discover" but will use this fact as a resource for generating those beliefs.[11] Nevertheless, it still might be thought that this association of objectivity with socially situated knowledge is an impossible combination. Has feminist standpoint theory really abandoned objectivity and embraced relativism? Or, alternatively, has it remained too firmly entrenched in a destructive objectivism that increasingly is criticized from many quarters?

Scientists and science theorists working in many different disciplinary and policy projects have objected to the conventional notion of a value-free, impartial, dispassionate objectivity that is supposed to guide scientific research and without which, according to conventional thought, one cannot separate justified belief from mere opinion, or real knowledge from mere claims to knowledge. From the perspective of this conventional notion of objectivity—sometimes referred to as "objectivism"—it has appeared that if one gives up this concept, the only alternative is not just a cultural relativism (the sociological assertion that what is thought to be a reasonable claim in one society or subculture is not thought to be so in another) but, worse, a judgmental or epistemological relativism that denies the possibility of any reasonable standards for adjudicating between competing claims. Some fear that to give up the possibility of one universally and eternally valid standard of judgment is perhaps even to be left with no way to argue rationally against the possibility that *each person's* judgment about the regularities of nature and their underlying causal tendencies must be regarded as equally valid. The reduction of the critic's position to such an absurdity provides a powerful incentive to question no further the conventional idea that objectivity requires value-neutrality. From the perspective of objectivism, judgmental relativism appears to be the only alternative.

Insistence on this division of epistemological stances between those that firmly support value-free objectivity and those that support judgmental relativism—a dichotomy that unfortunately has gained the consent of many critics of objectivism as well as its defenders—has succeeded in making value-free objectivity look much more attractive to natural and social scientists than it should. It also makes judgmental relativism appear far more progressive than it is. Some critics of the conventional notion of objectivity have openly welcomed judgmental relativism.[12] Others have been willing to tolerate it as the cost they think they must pay for admitting the practical ineffectualness, the proliferation of confusing conceptual contradictions, and the political regressiveness that follow from trying to achieve an objectivity that has been defined in terms of value-neutrality. But even if embracing judgmental relativism could make sense in anthropology and other social sciences, it appears absurd as an epistemological stance in physics or biology. What would it mean to assert that no reasonable standards can or could in principle be found for adjudicating between one culture's claim that the earth is flat and another culture's claim that the earth is round?

The literature on these topics from the 1970s and 1980s alone is huge and located in many disciplines. Prior to the 1960s the issue was primarily one of ethical and cultural absolutism versus relativism. It was the concern primarily of philosophers and anthropologists and was considered relevant only to the social sciences, not the natural sciences. But since then, the recognition has emerged that cognitive, scientific, and epistemic absolutism are both implicated in ethical and cultural issues and are also independently problematic. One incentive to the expansion was

Thomas Kuhn's account of how the natural sciences have developed in response to what scientists have found "interesting," together with the subsequent post-Kuhnian philosophy and social studies of the natural sciences.[13] Another has been the widely recognized failure of the social sciences to ground themselves in methods and theoretical commitments that can share in the scientificity of the natural sciences. Paradoxically, the more "scientific" social research becomes, the less objective it becomes.[14]

Further incentives have been such political tendencies as the U.S. civil rights movement, the rise of the women's movement, the decentering of the West and criticisms of Eurocentrism in international circles, and the increasing prominence within U.S. political and intellectual life of the voices of women and of African Americans and other people of Third World descent. From these perspectives, it appears increasingly arrogant for defenders of the West's intellectual traditions to continue to dismiss the scientific and epistemological stances of Others as caused mainly by biological inferiority, ignorance, underdevelopment, primitiveness, and the like. On the other hand, although diversity, pluralism, relativism, and difference have their valuable political and intellectual uses, embracing them resolves the political-scientific-epistemological conflict to almost no one's satisfaction.

I make no attempt here to summarize the arguments of these numerous and diverse writings.[15] My concern is more narrowly focused: to state as clearly as possible how issues of objectivity and relativism appear from the perspective of a feminist standpoint theory.

Feminist critics of science and the standpoint theorists especially have been interpreted as supporting either an excessive commitment to value-free objectivity or, alternatively, the abandonment of objectivity in favor of relativism. Because there are clear commitments within feminism to tell less partial and distorted stories about women, men, nature, and social relations, some critics have assumed that feminism must be committed to value-neutral objectivity. Like other feminists, however, the standpoint theorists have also criticized conventional sciences for their arrogance in assuming that they could tell one true story about a world that is out there, ready-made for their reporting, without listening to women's accounts or being aware that accounts of nature and social relations have been constructed within men's control of gender relations. Moreover, feminist thought and politics as a whole are continually revising the ways they bring women's voices and the perspectives from women's lives to knowledge-seeking, and they are full of conflicts between the claims made by different groups of feminists. How could feminists in good conscience do anything but abandon any agenda to a legitimate one over another of these perspectives? Many feminists in literature, the arts, and the humanities are even more resistant than those in the natural and social sciences to claims that feminist images or representations of the world hold any special epistemological or scientific status. Such policing of thought is exactly what they have objected to in criticizing the authority of their disciplinary canons on the rounds that such authority has had the effect of stifling the voices of marginalized groups. In ignoring these views, feminist epistemologists who are concerned with natural or social science agendas appear to support an epistemological divide between the sciences and humanities, a divide that feminism has elsewhere criticized.

The arguments [I have covered] move away from the fruitless and depressing choice between value-netural objectivity and judgmental relativism. [I previously] stressed the greater objectivity that can be and has been claimed to result from grounding research in women's lives. [I] draw on some assumptions underlying [earlier] analyses in order to argue that the conventional notion of objectivity against which feminist criticisms have been raised should be regarded as excessively weak. A feminist standpoint epistemology requires strengthened standards of objectivity. The standpoint epistemologies call for recognition of a historical or sociological or cultural relativism—but not for a judgmental or epistemological relativism. They call for the acknowledgment that all human beliefs—including our best scientific beliefs—are socially situated, but they also require a critical evaluation to determine which social situations tend to generate the most objective knowledge claims. They require, as judgmental relativism does not, a scientific account of the relationships between historically located belief and maximally objective belief. So they demand what I shall call *strong objectivity* in contrast to the weak objectivity of objectivism and its mirror-linked twin, judgmental relativism. This may appear to be circular reasoning—to call for scientifically examining the social location of scientific claims—but if so, it is at least not viciously circular.[16] . . .

The term "objectivism" is useful for the purposes of my argument because its echoes of "scientism" draw attention to ways in which the research prescriptions called for by a value-free objectivity only mimic the purported style of the most successful scientific practices without managing to produce their effects. Objectivism results only in semiscience when it turns away from the task of critically identifying all those broad, historical social desires, interests, and values that have shaped the agendas, contents, and results of the sciences much as they shape the rest of human affairs. Objectivism encourages only a partial and distorted explanation of why the great moments in the history of the natural and social sciences have occurred.

Let me be more precise in identifying the weaknesses of this notion. It has been conceptualized both too narrowly and too broadly to be able to accomplish the goals that its defenders claim it is intended to satisfy. Taken at face value it is ineffectively conceptualized, but this is what makes the sciences that adopt weak standards of objectivity so effective socially: objectivist justifications of science are useful to dominant groups that, consciously or not, do not really intend to "play fair" anyway. Its internally contradictory character gives it a kind of flexibility and adaptability that would be unavailable to a coherently characterized notion.

Consider, first, how objectivism operationalizes too narrowly the notion of maximizing objectivity. The conception of value-free, impartial, dispassionate research is supposed to direct the identification of all social values and their elimination from the results of research, yet it has been operationalized to identify and eliminate *only* those social values and interests that differ among the researchers and critics who are regarded by the scientific community as competent to make such judgments. If the community of "qualified" researchers and critics systematically excludes, for example, all African Americans and women of all races, and if the larger culture is stratified by race and gender and lacks powerful critiques of this stratification, it is not plausible to imagine that racist and sexist interests and values would be identified within a community of scientists composed entirely of people who benefit—intentionally or not—from institutional racism and sexism.

This kind of blindness is advanced by the conventional belief that the truly scientific part of knowledge-seeking—the part controlled by methods of research—is only in the context of justification. The context of discovery, where problems are identified as appropriate for scientific investigation, hypotheses are formulated, key concepts are defined—this part of the scientific process is thought to be unexaminable within science by rational methods. Thus "real science" is restricted to those processes controllable by methodological rules. The methods of science—or, rather, of the special sciences—are restricted to procedures for the testing of already formulated hypotheses. Untouched by these careful methods are those values and interests entrenched in the very statement of what problem is to be researched and in the concepts favored in the hypotheses that are to be tested. Recent histories of science are full of cases in which broad social assumptions stood little chance of identification or elimination through the very best research procedures of the day.[17] Thus objectivism operationalizes the notion of objectivity in much too narrow a way to permit the achievement of the value-free research that is supposed to be its outcome.

But objectivism also conceptualizes the desired value-neutrality of objectivity too broadly. Objectivists claim that objectivity requires the elimination of *all* social values and interests from the research process and the results of research. It is clear, however, that not all social values and interests have the same bad effects upon the results of research. Some have systematically generated less partial and distorted beliefs than others—or than purportedly value–free research.

Nor is this so outlandish an understanding of the history of science as objectivists frequently intimate. Setting the science for his study of nineteenth-century biological determinism, Stephen Jay Gould says:

> I do not intend to contrast evil determinists who stray from the path of scientific objectivity with enlightened antideterminists who approach data with an open mind and therefore see truth. Rather, I criticize the myth that science itself is an objective enterprise, done properly only when scientists can shuck the constraints of their culture and view the world as it really is. . . . Science, since people must do it, is a socially embedded activity. It progresses by hunch, vision, and intuition. Much of its change through time does not record a closer approach to absolute truth, but the alteration of cultural contexts that influence it so strongly.[18]

Other historians agree with Gould.[19] Modern science has again and again been reconstructed by a set of interests and values—distinctively Western, bourgeois, and patriarchal—which were originally formulated by a new social group that intentionally used the new sciences in their struggles against the Catholic Church and feudal state. These interests and values had both positive and negative consequences for the development of the sciences. Political and social interests are not "add-ons" to an otherwise transcendental science that is inherently indifferent to human society; scientific beliefs, practices, institutions, histories, and problematics are constituted in and through contemporary political and social projects, and always have been. It would be far more startling to discover a kind of human knowledge-seeking whose products could—alone among all human products—defy historical gravity" and fly off the earth, escaping entirely their historical location. Such a cultural phenomenon would be cause for scientific alarm; it would appear to defy principles of "material" causality upon which the possibility of scientific activity itself is based.[20] . . .

At this point, what I mean by a concept of strong objectivity should be clear. In an important sense, our cultures have agendas and make assumptions that we as individuals cannot easily detect. Theoretically unmediated experience, that aspect of a group's or an individual's experience in which cultural influences cannot be detected, functions as part of the evidence for scientific claims. Cultural agendas and assumptions are part of the background assumptions and auxiliary hypotheses that philosophers have identified. If the goal is to make available for critical scrutiny *all* the evidence marshaled for or against a scientific hypothesis, then this evidence, too, requires critical examination *within* scientific research processes. In other words, we can think of strong objectivity as extending the notion of scientific research to include systematic examination of such powerful background beliefs. It must do so in order to be competent at maximizing objectivity.

NOTES

1. Nancy Hartsock, "The Feminist Standpoint: Developing the Ground for a Specifically Feminist Historical Materialism," in *Discovering Reality,* ed. Sandra Harding and Merrill Hinitkka (Dordrecht, Holland: Reidel, 1983), p. 285. I have written about the distinction between feminist empiricist and feminist standpoint epistemologies in a number of places, usually in order to set the scene for a discussion of the less familiar standpoint epistemology. The most developed of such accounts are "Feminism and Theories of Scientific Knowledge," *American Philosophical Association Feminism and Philosophy Newsletter* 1 (1987); "Epistemological Questions," editor's conclusion to *Feminism and Methodology: Social Science Issues,* ed. Sandra Harding (Bloomington: Indiana University Press, 1987); and "Feminist Justificatory Strategies," in *Women, Knowledge, and Reality: Explorations in Feminist Philosophy,* ed. Ann Garry and Marilyn Pearsall (Boston: Unwin Hyman, 1989). The following analysis of why gender differences create scientific and epistemological resources is a slightly revised version of "Starting Thought from Women's Lives: Eight Resources for Maximizing Objectivity," *Journal of Social Philosophy* 21 (1990).

2. I have substituted "distorted" for "perverse," since one person's "perversities" may be another's most highly valued pleasures. "Distorted" appears less amenable to this kind of transvaluation.

3. In *The Science Question in Feminism,* I discussed differences

between the grounds proposed by four standpoint theorists: Hilary Rose, Nancy Hartsock, Jane Flax, and Dorothy Smith. Here I consider additional grounds proposed to justify feminist research.

4. Standpoint theories need not commit essentialism. *The Science Question in Feminism* contributed to such a misreading of their "logic"; [here] I contest an essentialist reading.

5. Jane Flax, "Political Philosophy and the Patriarchal Unconsciousness: A Psychoanalytic Perspective on Epistemology and Metaphysics," in *Discovering Reality,* ed. Harding and Minitkka. See also Nancy Hirschmann's use of object relations theory to ground a standpoint epistemology in her "Freedom, Recognition, and Obligation: A Feminist Approach to Political Theory," *American Political Science Review* 83, no. 4 (1989).

6. Sara Ruddick, *Maternal Thinking: Toward a Politics of Peace* (Boston: Beacon, 1989); Carol Gilligan, *In a Different Voice: Psychological Theory and Women's Development* (Cambridge: Harvard University Press, 1982); Mary Belenky et al., *Women's Ways of Knowing: The Development of Self, Voice, and Mind* (New York: Basic Books, 1986). I said above that Ruddick, Gilligan, and Belenky—among others—do not develop their criticisms of the generalization from stereotypically masculine to paradigmatically human reason into a standpoint epistemology. My point here is that their arguments can be used to do so.

7. Patricia Hill Collins, "Learning from the Outside Within: The Sociological Significance of Black Feminist Thought," *Social Problems* 33, no. 6 (1986).

8. Margaret Rossiter, *Women Scientists in America: 1940–1972* (Baltimore: Johns Hopkins University Press, 1995).

9. A good overview for modern science is Wolfgang van der Daele, "The Social Construction of Science," in *The Social Production of Scientific Knowledge,* ed. Everett Mendelsohn, Peter Weingart, and Richard Whitley (Dordrecht, Holland: Reidel, 1977).

10. See, for example, Boris Hessen, *The Economic Roots of Newton's Principia* (New York: Howard Fertig, 1971); Edgar Zilsel, "The Sociological Roots of Science," *American Journal of Sociology* 47 (1942). A historical precedent of a different sort is claimed by Marxist theorist Fredric Jameson, who argues that although it was Hungarian Marxist Georg Lukács who was responsible for the original development of standpoint theory, it is not Lukács's defenders today or other contemporary Marxists but feminist standpoint theorists who now exhibit "the most authentic

descendency of Lukács's thinking." See Fredric Jameson, "History and Class Consciousness as an 'Unfinished Project,'" in *Rethinking Marxism* 2, no. 1 (1998): 49–72; Lukács, *History and Class Consciousness* (Cambridge: MIT Press, 1971).

11. See Donna Haraway, "Situated Knowledge: The Science Question in Feminism and the Privilege of Partial Perspective," *Feminist Studies* 14, no. 3 (1988).

12. For example, David Bloor, *Knowledge and Social Imagery* (London: Routledge & Kegan Paul, 1977); and many of the papers in *Knowledge and Reflexivity,* ed. Steve Woolgar (Beverly Hills: Sage, 1988).

13. Thomas Kuhn, *The Structure of Scientific Revolutions* (Chicago: University of Chicago Press, 1970).

14. This is an important theme in Richard Bernstein, *Beyond Objectivism and Relativism* (Philadelphia: University of Pennsylvania Press, 1983). Similar doubts about the ability of legal notions of objectivity to advance justice appear in many of the essays in "Women in Legal Education: Pedagogy, Law, Theory, and Practice," *Journal of Legal Education* 38 (1988), special issue, ed. Carrie Menkel-Meadow, Martha Minow, and David Vernon.

15. Discussions on one or more of these focuses can be found in Martin Hollis and Steven Lukes, eds., *Rationality and Relativism* (Cambridge: Harvard University Press, 1982); Michael Krausz and Jack Meiland, eds. *Relativism: Cognitive and Moral* (Notre Dame: University of Notre Dame Press, 1982); Richard Bernstein, *Beyond Objectivism;* and S. P. Mohanty, "Us and Them: On the Philosophical Bases of Political Criticism," *Yale Journal of Criticism* 2, no. 2 (1989). A good brief bibliographic essay on the recent philosophy of science within and against which the particular discussion of this chapter is located is Steve Fuller, "The Philosophy of Science Since Kuhn: Readings on the Revolution That Has Yet to Come," *Choice* (December 1989). For more extended studies that are not incompatible with my arguments here, see Steve fuller, *Social Epistemology* (Bloomington: Indiana University Press, 1988); and Joseph Rouse, *Knowledge and Power: Toward a Political Philosophy of Science* (Ithaca: Cornell University Press, 1987).

16. Additional writings informing this chapter include especially Haraway, "Situated Knowledges"; Haraway, *Primate Visions: Gender, Race, and Nature in the World of Modern Science* (New York: Routledge, 1989); Jane Flax, *Thinking Fragments: Psychoanalysis, Feminism, and Postmodernism in the Contemporary West* (Berkeley: University of California Press, 1990);

and the writings of standpoint theorists themselves, especially Nancy Hartsock, "The Feminist Standpoint: Developing the Ground for a Specifically Feminist Historical Materialism," in *Discovering Reality: Feminist Perspectives on Epistemology, Metaphysics, Methodology, and Philosophy of Science,* ed. Sandra Harding and Merrill Hintikka (Dordrecht: Reidel, 1983); Dorothy Smith, *The Everyday World as Problematic: A Feminist Sociology* (Boston: Northeastern University Press, 1987); Hilary Rose, "Hand, Brain, and Heart: A Feminist Epistemology for the Natural Sciences," *Signs* 9, no. 1 (1983); Patricia Hill Collins, "Learning from the Outsider Within: The Sociological Significance of Black Feminist Thought"; though each of these theorists would no doubt disagree with various aspects of my argument.

17. This is the theme of many feminist, left, and antiracist analyses of biology and social sciences. See, for example, Anne Fausto-Sterling, *Myths of Gender: Biological Theories about Women and Men* (New York: Basic Books, 1985); Stephen Jay Gould, *The Mismeasure of Man* (New York: Norton, 1981); Robert V. Guthrie, *Even the Rat Was White: A Historical View of Psychology* (New York: Harper & Row, 1976); Haraway, *Primate Visions*; Sandra Harding, ed., *Feminism and Methodology: Social Science Issues* (Bloomington: Indiana University Press, 1987); Joyce Ladner, ed., *The Death of White Sociology* (New York: Random House, 1973); Hilary Rose and Steven Rose, eds., *Ideology of/in the Natural Sciences* (Cambridge, Mass.: Schenkman, 1979); Londa Schiebinger, *The Mind Has No Sex: Women in the Origins of Modern Science* (Cambridge: Harvard University Press, 1989).

18. Gould, *Mismeasure of Man,* pp. 21–22.

19. For example, William Leiss, *The Domination of Nature* (Boston: Beacon Press, 1972); Carolyn Merchant, *The Death of Nature: Women, Ecology, and the Scientific Revolution* (New York: Harper & Row, 1980); Wolfgang van den Daele, "The Social Construction of Science," in *The Social Production of Scientific Knowledge,* ed. Everett Mendelsohn, Peter Weingart, and Richard Whitley (Dordrecht: Reidel, 1977).

20. See Rouse, *Knowledge and Power,* chap. 4, which provides a good analysis of the implications for science of Foucauldian notions of politics and power.

SIX

FEMINIST EPISTEMOLOGY

Implications for Philosophy of Science

CASSANDRA L. PINNICK

The reason the feminist claims can turn out to be scientifically preferable is that they originate in, and are tested against, more complete and less distorting kinds of social experience. The experiences arising from the activities assigned to women, seen through feminist theory, provide a grounding for potentially more complete and less distorted knowledge claims than do men's experi-

Reprinted from Cassandra L. Pinnick, "Feminist Epistemology: Implications for Philosophy of Science," *Philosophy of Science* 61 (1994): 646–57.

ences. This kind of politicized inquiry increases the objectivity of the results of research.

–Sandra Harding, "Feminist Justificatory Strategies"

INTRODUCTION

The central thesis of this article is that *feminist* epistemology should not be taken seriously. This is because any feminist epistemology which radically challenges traditional theories of knowledge is unable to resolve the tension between (a) its thesis that every epistemology is a sociopolitical artifact, and (b) its stated aim to articulate an epistemology that can be *justified* as better than its rivals.

To develop these issues, I concentrate on the influential work of S. Harding. Harding builds upon larger efforts to articulate a feminist perspective on society, culture, politics, and economics. She presents the strongest case for an epistemologically relativist, feminist critique of science, using various interpretations of T. Kuhn's *The Structure of Scientific Revolutions*[1] and W. V. O. Quine's underdetermination thesis, the Strong Program in the sociology of scientific knowledge, and general themes within the feminist critique of modern society. Her writings represent a forceful expansion of feminist theory into well-developed and mature areas of epistemology, and her works are cited widely in cognate fields, especially in the social studies of science.

Harding argues that feminists as epistemologists, as philosophers of science, and as scientists can and should improve science. I focus on Harding's epistemic claims on behalf of a feminist epistemology of science. Indeed, when read carefully, Harding says nothing about the plight of women generally; her arguments reach only to the fate of feminists,[2] and she clearly takes the scope of "woman" to be distinct from that of "feminist."[3] This differentiates her focus—at least when she writes as a philosopher of science—from that of traditional feminism which has interesting things to say about the political status of women.

I take epistemology of science to be concerned with questions about the nature of evidence for or against scientific beliefs, and with the critical assessment of the presuppositions and arguments of rival theories of scientific knowledge. One way to carry out this task is to look for indicators that a particular epistemology distinguishes itself from rivals. An epistemology might do this (1) by resolving traditional problems that have confounded other epistemologies, (2) by disclosing important new problems that have been overlooked or addressed in less-than-satisfactory means by its rivals, and (3) by using better methods to realize stated scientific aims. These epistemic benchmarks have special bearing on my critique of Harding's provocative theses about science. I focus on her controversial *epistemological* thesis that feminist and other "liberationist" theories of knowledge provide the only uncorrupted *objective* method for the evaluation of scientific claims.

FEMINIST EPISTEMOLOGY AND EMPIRICAL METHODS

Some feminist arguments attack the empirical method which is thought to provide science's epistemic rationale, unlike feminist critiques of science that focus on particular instances of sexual bias in science. Harding's criticisms, if correct, would demonstrate that empiricist epistemology and philosophy of science fail to live up to traditional empiricist standards of objective inquiry, driven by universalizable cognitive norms. She tries to show that science is an irretrievably male-biased tool of a sociopolitical power elite.

The problem with science, as Harding sees it, is not sexism. Instead, the problem is that scientific knowledge reflects a set of noncognitive interests and values which serve the political ends of Western-European, white males, while suppressing other social groups, "[Men] are a particularly poor grounding for knowledge claims since, as masculine, they represent the ruling part of society";[4] "[S]cience is just one way of perpetrating and legitimating male dominance."[5] Those in control of science are concerned with maintaining political power and with "obscuring the injustices of their unearned privileges and authority,[6] thus the democratic ideal of science-for-all is impossible under the true conditions that motivate science.

Harding's remedy is not to strive for more diligence in rooting out intrusive political influences. Instead, she claims that only when the political influences that control science are acknowledged can scientific inquiry achieve genuinely objective results. For this reason feminist epistemology "sets the relationship between knowledge and politics at the center of its account in the sense that it tries to provide causal accounts to explain the effects that different kinds of politics have on the production of knowledge."[7] A feminist science can be genuinely objective because feminists' political status gives them a special vantage point from which to discharge the aims of science.[8] Although all epistemological perspectives distort the true nature of reality, Harding states that a feminist perspective is less distorting than others.

Despite its tone and reductionist tendencies,[9] Harding's work does not belong to the science-bashing genre. A consistent thread in her writing argues for an improved science, not for its elimination. Her arguments on behalf of radical epistemological departures are based on the promise that a fundamental restructuring of its means will improve science. In particular, Harding argues that breaking the traditional identification of objectivity with neutrality or disinterestedness will result in better insight on nature and a concomitant improved capacity to do science. The neutrality ideal subverts scientific aims because it "defends and legitimates the institutions and practices through which powerful groups can gain the information and explanations that they need to advance their priorities."[10]

Harding's scheme for feminist epistemology of science yields the surprising consequence that the *less* politically neutral the basis and conduct of scientific inquiry, the *more* objective the results, an anathema for most of us who are familiar with instances of politically motivated science, such as Shockley's eugenics, or Brigham's and Grant's aptitude and intelligence-test designs. But surprising or counterintuitive theoretical consequences alone neither prove nor provide compelling grounds to suspect that Harding's arguments are wrong. Such consequences do prove the ambitious nature of her brief against the methodology of traditional empirical science. If correct, she forces a basic restructuring of empiricist epistemology and philosophy of science. Unfortunately, her arguments fail for the reason either that they rely on contested and dubious philosophical positions or that they lack data to support the interesting empirical claims that she advances.

FEMINIST STANDPOINT EPISTEMOLOGY

A tradition in the philosophy and history of science holds that objective, politically neutral inquiry maximizes the power to achieve scientific aims such as devising theories that are good predictors of natural phenomena over long periods of time for the kind of phenomena which they are designed to describe. Objectivity may be an ideal case, but despite shortfalls, historical evidence apparently supports its epistemological worth. Harding, in contrast, argues that objectivity in scientific research is a delusion, and as traditionally understood, no boon to science: "[T]he problem with the conventional conception of objectivity is not that it is too rigorous or too 'objectifying,' as some have argued, but that it is *not rigorous or objectifying enough:* it is too weak to accomplish even the goals for which it has been designed, let alone the more difficult projects called for by feminisms and other new social movements."[11] To begin to comprehend Harding's novel claim, I review how her recent thinking on feminist epistemology bears on the role she assigns to objectivity in a feminist philosophy of science.

Harding's favored species of feminist epistemology is what she calls "feminist standpoint epistemology." Two basic claims underlie the theory. First, empiricist epistemology is based on the utopian ideal of objective inquiry that, in fact and in principle, impedes scientific progress. Thus, science cannot and should not strive to live up to the stated standards of empiricist epistemology, and feminist standpoint theorists reject the notion of disinterested, value-free, *objective,* scientific inquiry:

> The feminist standpoint, like feminist empiricism, clearly asserts that objectivity never has been and could not be increased by the exclusion or elimination of social values from inquiry. . . . [I]t is commitment to anti-authoritarianism, anti-elitism, and anti-domination tendencies that has increased the objectivity of science and will continue to do so.[12]

This point is not argued successfully. First, Harding fails to show that we cannot "socialize" epistemology, but retain the concept of objective standards and rational inquiry that have been central to an empiricist theory of knowledge.[13] Few philosophers of science presently deny that noncognitive factors play a role in science; yet, this concession to the effect of noncognitive influences on scientific belief does not endorse the slide to an arational account of science.

Harding's second contention—that feminists, being a "marginalized" social group, offer a better perspective on which to base scientific inquiry—is more interesting.[14] She maintains that scientific results based on the perspective of marginalized persons, such as feminists, better represent nature and more nearly achieve a democratic ideal of knowledge than do scientific results based on male-oriented practices. Thus, feminist perspectives on nature are, in the true sense of the term, *objective* results: "Standpoint theory provides resources for the stronger, more competent standards for maximizing objectivity that can advance our abilities to distinguish between how different social groups want the world to be and how 'in fact' it is."[15]

Harding calls objectivity based on politically guided scientific inquiry "strong objectivity."[16] This claim stands at the heart of feminist standpoint epistemology. If Harding presents—as she promises evidence that feminists, as marginalized persons or as a marginalized social group, do science better than nonmarginalized persons, she can show (1) that objectivity, a fundamental concept of traditional empiricist epistemology, must be redefined; and (2) that certain types of politically situated persons should be at the reins of science.

PHILOSOPHICAL INDUCEMENTS FOR A STANDPOINT EPISTEMOLOGY

Before discussing the epistemic merits of strong objectivity, I discuss the philosophical impetus behind standpoint epistemology. Harding[17] motivates feminist standpoint epistemology in several ways, but she grounds the theory on an interpretation of arguments that she and many sociologists of science attribute to Kuhn and Quine.

Whatever Kuhn's and Quine's intentions, feminist epistemologists, and their programmatic fellow travelers, clearly see Kuhn and Quine as having clinched the case for an arational, sociopolitical analysis of science. For example, Harding writes that "in effect, [Kuhn] showed that all of natural science was located inside social history. . . . [A]ny theory can always be retained as long as its defenders hold enough institutional power to explain away potential threats to it."[18] For feminist standpoint epistemology, the noncognitive interests of Western European, white males, who dominate and control science, fill the putative gap opened by Kuhn's and Quine's analyses of the epistemological foundations of scientific knowledge.

Harding's reliance on this interpretation speaks only to the already converted. If she wants to rework science from the inside out,[19] then she needs arguments that will draw more than a yawn from philosophers of science who have expended considerable effort voicing objections to this use of the Kuhn-Quine corpus. Specifically, philosophers of science deny that the combined works of Kuhn and Quine license, even less necessitate, arational analysis of science, for the reason that no one has yet shown that admitted logical gaps in scientific reason must be filled by noncognitive, sociopolitical, that is, *arational*, causal explainers.[20] Nor has anyone demonstrated the plausibility, much less the truth, of the existence claim that any number of possible interpretations are equally warranted under the conditions of a particular experimental project or environment.[21] Without answers to deflect these (and other) philosophical objections to the sort of use to which she puts Kuhn's and

Quine's work, Harding's feminist theory of science is gratuitous, and the disinclined need pay it no attention.

Further inspiration mentioned by Harding for standpoint epistemology is an underlying intellectual debt to Marxist theory,[22] and intellectual ties to the Strong Program in the sociology of scientific knowledge.[23] But, Harding notes, neither Marxist political theory nor Strong Program sociology of science are sufficiently radical, "[T]he standpoint theorists see gender relations as at least as causal as economic relations in creating forms of social life and belief. . . . In contrast to Marxism, women and men are not merely (or perhaps even primarily) members of economic classes."[24] She excuses Marxism on the grounds that it was not historically situated to be a successor epistemology. But she chides the Strong Program for not seeing that gender issues need to be taken into account to fill out the program's analysis—hence the need to out-macho the Strong Program with her demand for "strong" objectivity.[25]

Harding does, however, embrace the Strong Program as a source of convincing historical case studies which reveal the political nature of science that underwrites the feminist standpoint challenge.[26] Here again, Harding relies on controversial evidence to support her call for drastic epistemological change. The force of Strong Program case studies, especially their success in establishing the conclusion that philosophical accounts of scientific change can and should be replaced by arational accounts, has drawn sharp criticism. No Strong Program case study successfully reduces science to politics, and in certain key instances the historical scholarship is selectively focused.[27]

However, even without questioning the historical reliability of Strong Program case studies, or the putative advantages attributed to the Strong Program's reductionistic program, Harding finds no philosophical grounding here to motivate feminist standpoint epistemology. No Strong Program case study shows *causal* connections between scientific belief and concomitant sociopolitical allegiances. At best, these studies establish temporal coincidence between cognitive and noncognitive commitments.[28] Still, even if some historical pattern of such temporal coincidence were demonstrated, the Strong Program sample is too small to support generalization. In any event, none of the case studies allows for any familiar type of empirical control.

This leads to the final problem with Harding's appeal to the Strong Program. The underlying methodology of this brand of sociological analysis of scientific change is fatally defective because each case study relies on counterfactual reasoning. And, in this instance at least, counterfactual reasoning is not persuasive because it is impossible within the venue of an historical case study to gather the necessary inductive evidence that could favor a particular explanation of the actual events over some other, allegedly possible, set of events or historical outcomes. Effective inductive regularities of the kind needed to support Strong Program case studies are not of the following type: *In similar instances of scientific change observed in the past, certain types of cognitive beliefs, Y, have been regularly associated with certain types of sociopolitical allegiances, X. Thus, we may hold that in the present circumstances, the presence of X-type of political alliances signals the presence of Y-type cognitive beliefs.* If demonstrated by the Strong Program's case studies, this kind of regularity possibly links cognitive beliefs to noncognitive causal factors (but, importantly, the causal issue would remain open). However, these kinds of regularities do nothing to establish the plausibility of the historical accounts that the Strong Program requires.

The Strong Program needs to show that the historical record of science, so far as *rational* considerations are concerned, could always be different than it actually is. So, for example, had different political forces triumphed, some form of a priorism might have held sway in seventeenth-century England rather than experimentalism. Now, if the Strong Program argument is something more than the (obviously false) claim that the imaginability of a different outcome implies the genuine possibility that the outcome could be different, then inductive evidence of the following kind is needed: *In the past, it has been observed that similar cases of scientific change have been resolved differently than they actually were resolved.* However, it is impossible to observe or compare actual history with alternative histories, and the Strong Program can only tell just-so stories about how science might have been.

The Strong Program's deep problems infect Harding's analysis. The Strong Program and Harding each contend that complete plasticity of cognitive factors, in every case, supports the conclusion that particular instances of scientific change could have been resolved differently than they in fact were—even contradictory to the actual course of history; and that, furthermore, according to the Strong Program argument, explaining the actual historical record of science necessarily reduces to an interplay of noncognitive causal factors. However, as I indicate, the kind of evidence required to substantiate Strong Program reductionist claims is precisely what the Strong Program seeks to prove, so that the Strong Program's reliance on the historical record of science amounts to no more than a *petitio*. As such, Harding's cause is poorly served.

MARGINALIZED PERSONS AND EPISTEMIC PRIVILEGE

I now return to Harding's[29] argument that because feminists are marginalized, they have a privileged perspective on nature. Although a marginalized perspective on nature is not infallible, it does provide a less distorted view than that from within the dominant group. This appears to be a good empirical claim, open to evaluation based on empirical data. One expects that Harding will turn her efforts to show that marginalized feminists have either a record of obtaining better results than nonfeminists and other nonmarginalized types, or that a small but remarkable body of data (inconclusive though it may be at the present) suggests that marginalized feminists could more successfully achieve scientific ends. The comparative success rates could be evaluated with regard to certain practical applications in, for example, the sciences of agriculture, medicine, or engineering. In the spirit of traditional empiricist philosophy of science, Harding's claim on behalf of feminist epistemic privilege has the welcome potential to move the discussion from an exchange of favored a priori, philosophical arguments to the relative merits of competing empirical claims.

However, this literature describes no effort to accumulate the kind of empirical data that could easily resolve matters in favor of the feminists. Philosophers of science have acknowledged the need for data in the face of challenges that seemed to come from Kuhn and Quine;[30] in their best interests, feminists should make a similar bow. To date, feminist standpoint epistemology offers no data to support the epistemological advice that marginalized persons should take the place of present scientists in the ranks and at the cutting edge of science.

To be fair to Harding's argument, she states that "historical precedents" establish that marginalized people are at the truly progressive frontiers of scientific change.[31] To co-opt her terminology, this is an instance of "bad-Kuhn" in her theorizing. The claim that only marginalized persons can effect change echoes Planck's hypothesis which says that scientific change must wait for older scientists, those most entrenched in present scientific thought, to die off and be replaced by a new generation of thinkers who are less blinded to change and have no stake in maintaining the intellectual status quo.[32]

Rightly or wrongly, many thinkers (especially in the social sciences) regard this as an important truism. Harding joins ranks with present-day philosophers, historians, and sociologists who agree that age and entrenchment negatively affect the readiness with which scientists change their minds.[33] Enculturated minds might be more difficult to change than unformed minds. Mature scientists in the center of things presumably should be more committed to received views than beginners at the periphery of scientific circles.

These conventional truths suggest that the future of science rests with those who have a fresh approach—young scientists at the margins of scientific power—but do not require us to turn science over to, for example, marginalized feminists.

Furthermore, empirical data discredits the intuition underlying the Planck hypothesis. For example, Hull et al.[34] test the Planck hypothesis against a particular episode in the history of science. The results establish that the connection between age or membership in a scientific elite and acceptance of a new scientific idea by those on the fringes of science is less important than Planck claimed. Indeed the statistical results indicate that if age correlates with an entrenched, nonmarginalized position of power in science, then older scientists and marginal, younger scientists adopt new scientific concepts at a similar rate.

Of course, this study and others like it do not foreclose the possibility that feminists have a privileged epistemological view on which science should be based. But if Harding and other standpoint feminist epistemologists intend their arguments to be taken seriously outside their own circles, then they must direct their efforts to designing studies that will generate data suggesting feminists do better science. Specifically, Harding needs to show that politically motivated research, under the guidance of feminists, accomplishes scientific aims better than research done under the auspices of the traditional empiricist, socially and politically disengaged, ideal inquirer. The empirical nature of Harding's claim on behalf of a feminist restructuring of science requires data to do the showing. At present, no data for any component of this thesis is cited in Harding's work.

FEMINIST METHODS TO MAXIMIZE OBJECTIVITY

Before concluding that feminist standpoint epistemology has nothing new and interesting to bring to an epistemology of science, I want to address three peculiarly feminist methods for maximizing objectivity that Harding[35] summarizes.

First, issues important to women's lives, "distinctive features of women's social situations,"[36] have been overlooked in the course of scientific inquiry, and feminist scientists will affect the content of scientific research. However, this obvious truth does not demand radically restructuring empiricist epistemology. Indeed, one of feminism's strongest and most positive intellectual influences arguably has been in areas of scientific research that were long ignored but now command attention. Second, marginalized feminists have less to lose, and so they will be more inclined to question accepted scientific beliefs that need closer scrutiny.[37] As I have argued, despite its intuitive appeal, this is a variant of the unsuccessful Planck hypothesis. Third, feminist standpoint epistemology is historically appropriate for this time.[38] However, no evidence supports this *ad populum* claim.

CLOSING REMARKS

I have concentrated primarily on the lack of sound philosophical argument or empirical support for the most daring feminist epistemological proposals. The lack of empirical support is disappointing and damaging at present to the prospects for a feminist epistemology and philosophy of science. However, certain flagrant philosophical dilemmas cannot be ignored entirely.

It must be noted, first, that if Harding is correct that feminists are marginalized, and if it is correct that marginalization confers epistemic privilege, one wonders what happens when and if feminists achieve their goals. The standpoint case for feminist science hinges on the claim that feminists, by virtue of being a repressed political minority, acquire a special insight into the nature of natural processes. This is a blatant non sequitur. But, even worse, by this very argument, should feminists achieve political equality, they would thereby lose any claim to epistemic privilege, and feminist science would accordingly lose its claim to superiority over nonfeminist science.

Also, if Harding chooses to use the philosophical arguments that she believes license a standpoint theory of knowledge, arguments relying on Kuhn and Quine and theorizing associated with the Strong Program, then she must own up to the logical consequences of such views. Thus, it becomes inconsistent for her to say, on the one hand, that every epistemology is a tool of the power elite and at the same time maintain that a particular epistemology, feminist standpoint, will generate "less distorted" methods and beliefs. The first claim forecloses the possibility of justifying the latter type of claim on behalf of any particular epistemology.

This problem is compounded when Harding's argument expands, as it does, to include "multiple perspectives."[39] As she says, each of these "liberationist epistemologies" is credible. The "logic" of liberationist epistemologies "leads to the recognition that the subject of liberatory feminist knowledge must also be, in an important if controversial sense, the subject of every other liberatory knowledge project."[40] But, what kind of advice does feminist standpoint epistemology have when the perceptions of different liberationist epistemologies conflict? None. And Harding views this as a welcome consequence, "In the contradictory nature of this project lies both its greatest challenge and a source of its great creativity."[41]

A philosophy of science qua social science whose only goal is to tell inconsistent and incoherent stories is not very appealing or sufficiently ambitious.

NOTES

1. Thomas Kuhn, *The Structure of Scientific Revolutions*, 2d ed. (Chicago: University of Chicago Press, 1970).

2. Sandra Harding "Feminist Justificatory Strategies," in *Women, Knowledge, and Reality*, ed. A. Garry and M. Pearsall (Boston: Unwin Hyman, 1989), p. 197.

3. Sandra Harding, "Rethinking Standpoint Epistemology: What Is 'Strong Objectivity,'" *Centennial Review* 36 (1992): 457.

4. Sandra Harding, "How the Women's Movement Benefits Science: Two Views," *Women's Studies International Forum* 12 (1989): 274.

5. Ibid., p. 281.

6. Ibid.

7. Harding, "Rethinking Standpoint Epistemology," p. 444.

8. Harding, "How the Women's Movement Benefits Science," p. 274.

9. See, for example, Sandra Harding, "Women as Creators of Knowledge," *American Behavioral Scientist* 32 (1989): 700.

10. Sandra Harding, "After the Neutrality Ideal: Science, Politics, and 'Strong Objectivity,'" *Social Research* 59 (1992): 568.

11. Harding, "Rethinking Standpoint Epistemology," p. 438.

12. Harding, "Feminist Justificatory Strategies." Harding claims that

"feminist empiricism" puts us on the alert for what she calls "bad science," but is powerless to set in place safeguards that will prevent repeated instances of the male-biased practices and policies which pervert the spirit of scientific inquiry (Harding, "How the Women's Movement Benefits Science," pp. 281–82).

13. For discussion of this possibility, see L. Laudan "Demystifying Underdetermination," *Minnestoa Studies in the Philosophy of Science* 14 (1990): 267–97; P. Kitcher, "The Division of Cognitive Labor," *Journal of Philosophy* 3 (1990): 5–22; and A. Goldman, "Foundations of Social Epistemics," *Synthese* 73 (1987): 109–44.

14. Harding, "How the Women's Movement Benefits Science," p. 274. Harding does not want to rule out that men, as well as women, may turn into feminists, and thus, presumably, take on an epistemically privileged vantage point to criticize science (see p. 281). For this reason, I take her to intend that *feminists* belong to the class of marginalized persons, and that feminists are all those who see "through feminist theory," not just women.

15. Sandra Harding, "Starting Thought from Women's Lives: Eight Resources of Maximizing Objectivity," *Journal of Social Philosophy* 21 (1990): 140–49.

16. Ibid.; Harding, "Rethinking Standpoint Epistemology."

17. Harding, "After the Neutrality Ideal," p. 582, n. 13.

18. Ibid.; see also, Harding, "Rethinking Standpoint Epistemology," p. 440.

19. Harding, "Starting Thought from Women's Lives," p. 146.

20. See Laudan, "Demystifying Underdetermination"; P. Slezak, "Bloor's Bluff: Behaviorism and the Strong Programme," *International Studies in the Philosophy of Science* 5 (1991): 241–56.

21. For details of this particular criticism against appropriating Quine's work to the relativists' cause, see L. Laudan and L. Lepin, "Empirical Equivalence and Underdetermination," *Journal of Philosophy* 88 (1991): 449–90.

22. Harding, "Starting Thought from Women's Lives," p. 140.

23. Harding, "Rethinking Standpoint Epistemology," p. 463. For discussion see Bloor, *Knowledge and Social Imagery.*

24. Harding, "Feminist Justificatory Strategies," p. 197.

25. Harding, "Starting Thought from Women's Lives," p. 146; Harding, "Rethinking Standpoint Epistemology," p. 463.

26. Harding, "Rethinking Standpoint Epistemology," p. 460.

27. See, for example, P. Roth and R. Barrett, "Deconstructing Quarks," *Social Studies of Science* 20 (1990): 579–623.

28. Ibid.

29. Sandra Harding, "Subjectivity, Experience, and Knowledge: An Epistemology from/for Rainbow Coalition Politics," *Development and Change* 23 (1992): 186, 189.

30. R. Laudan, L. Laudan, and A. Donovan, *Scrutinizing Science: Empirical Studies of Scientific Change* (Dordrecht, Holland: Kluwer, 1988).

31. Harding, "How the Women's Movement Benefits Science," p. 280.

32. Planck had not been the only thinker in modern science to express this sentiment. Lavoisier, for example, remarked that "[t]he human mind gets creased into a way of seeing thing. Those who have envisaged nature according to a certain point of view during much of their career, rise only with difficulty to new ideas" (quoted in D. Hull, P. Tessner, and A. Diamond, "Planck's Principle," *Science* 202 [1978]: 717). And, English biologist T. H. Huxley is notable for having advised that men of science ought to be strangled on their sixtieth birthday "lest age should harden them against the reception of new truths, and make them into clogs upon progress, the worse, in proportion to the influence they had deservedly won" (quoted in L. Huxley, *Life and Letters of Thomas Henry Huxley,* vol. 2 [New York: Appleton, 1901], p. 117).

33. For example, see Kuhn, *The Structure of Scientific Revolutions,* p. 151; P. Feyerabend, "Consolations for the Specialist," in *Criticism and the Growth of Knowledge,* ed. O. Lakatos and A. Musgrave (Cambridge: Cambridge University Press, 1970), p. 203. Both Kuhn and Feyerabend quote Plank's principle in support of their thesis that scientific change is, at bottom, arational.

34. Hull et al., "Planck's Principle."

35. Harding, "Starting Thought from Women's Lives."

36. Ibid., p. 140.

37. Ibid., p. 145.

38. Ibid., p. 146.

39. See discussion in Harding, "Rethinking Standpoint Epistemology."

40. Ibid., p. 455.

41. Ibid., p. 448.

SEVEN

A CRITIQUE OF HARDING ON OBJECTIVITY

ELLEN R. KLEIN

OBJECTIVITY UNDER SIEGE

The concept of "objectivity" is articulated in a number of ways by a number of different feminists working in science criticism. For Ruth Bleier and Catharine MacKinnon, for example, "objectivity" is synonymous with a "value-free stance,"[1] and the "noninvolved stance,"[2] respectively. Evelyn Fox Keller

Reprinted from Ellen R. Klein, "A Feminist Crtitique of Science," in *Feminism Under Fire* (Amherst, N.Y.: Prometheus Books, 1996), pp. 36–50.

claims the objectivist ideology proclaims "disinterest,"[3] a formulation similar to Jean Grimshaw's understanding of "objectivity" as "impartiality."[4]

Some feminists go so far as to claim the desire for "objectivity" is merely male fantasy. Susan Bordo,[5] for instance, argues that the Cartesian desire/obsession for objectivity is an expression of anxiety over separation from the organic female universe, and thus constitutes an intellectual flight from the feminine, and a desire for control.[6] In general, then, "objectivity" is simply "pernicious."[7]

The Two Faces of Harding's Conception of Objectivity

Sandra Harding understands the concept of "objectivity" in two disparate ways. The first is in terms of what has been known in the literature as the traditional notion of "objectivity," which, as stated by a number of feminists (above), is synonymous with an attempt to achieve "value-free, impartial, dispassionate research."[8] The second is less conventional and, interestingly, at odds with the first.

Via a "feminist standpoint"[9] Harding has "appropriate[d] and redefine[d] objectivity,"[10] claiming that objectivity "is not maximized through value-neutrality."[11] Instead, she claims that "feminist standpoint theory can direct the production of less partial and less distorted beliefs."[12] The former traditional notion of objectivity she calls "weak objectivity"; the latter she calls "strong objectivity."

The traditional notion is pegged as "weak," for it conceptualizes value-neutrality in a way that is at once too narrow and too broad. It is too narrow, according to Harding, because it tends to identify and eliminate "*only* those social values and interests that differ among the researchers and critics who are regarded by the scientific community as competent to make such judgments."[13] Such selective elimination of bias is, itself, biased. She fantasizes:

> If the community of "qualified" researchers and critics systematically excludes, for example, all African Americans and women of all races, and if the larger culture is stratified by race and gender and lacks powerful critiques of this stratification, it is not plau-

> sible to imagine that racist and sexist interests and values would be identified within a community of scientists composed entirely of people who benefit—intentionally or not—from institutional racism and sexism.[14]

The traditional, i.e., "weak," concept of objectivity is, according to Harding, also too broad,

> for it requires the elimination of all social values and interests from the research process and the results of research. It is clear, however, that not all social values and interests have the same bad effects upon the results of research. Some have systematically generated less partial and distorted beliefs than others.[15]

Harding suggests that feminist values are one such case.

Because of the above, Harding recommends that scientists (and philosophers of science) stop looking for "weak" objectivity; instead, they should concentrate on trying to achieve the "strong" variety. The muscular notion of objectivity extends "the notion of scientific research to include [the] systematic examination"[16] of one's "cultural influences . . . cultural agendas . . . background assumptions . . . auxiliary assumptions . . . [and even] the macro tendencies in the social order."[17]

In the final analysis, "weak" objectivity is represented by the traditional notion—value-free, noninvolved, impartial, and/or disinterested stance—expressed in one or all of these ways by any number of philosophers throughout the years. "Strong" objectivity—the emancipatory variety—is vintage Harding.

Interpretations of Objectivity

In what follows I will examine both the "weak" and "strong" notions of "objectivity." I will show that the "strong" account is too obviously biased and political to take seriously as a target for criticism from either traditional or feminist camps. I will show that the "weak" concept, fleshed out in terms of value-free or noninvolved stance—what I will call "weak no. 1"—caricatures the traditional concept of objectivity held by most scientists and philosophers of science and therefore need not be defended against feminist criticism. The "weak" concept of objectivity, expressed in terms of impartiality or disinterest—what I will call "weak no. 2"—accurately reflects the sentiments of most scientists and philosophers of science; but the criticisms leveled against this interpretation are not sufficient to support the claim that science is itself sexist.

Objectivity as Emancipatory

Harding's attempt to articulate an account of "strong objectivity" is most salient in the following passage: "The paradigm models of objective science are those studies explicitly directed by morally and politically emancipatory interests—that is, by interests in eliminating sexist, racist, classicist, and culturally coercive understandings of nature and social life."[18] This attempt, I believe, has already been thoroughly criticized by Kristin Shrader-Frechette.[19] Therefore, I will only point out the relevant passage in her critique of Harding. Schrader-Frechette states that because

> Harding is not employing the term "objectivity" in its ordinary sense . . . her use is question-begging both because she has not defined it, and because this sense of the term is highly stipulative . . . [that is] she does not explain how scientific work becomes more objective by being directed by moral and political *interests* . . . how work expressing moral and political *values* lays claim to objectivity.[20]

Since this critique, however, Harding has attempted to alter her response to meet this objection. In her more recent article, "Starting from Women's Lives," Harding claims that "maximizing objectivity requires critically examining not only those beliefs that differ between individuals . . . but also those that are held by virtually everyone who gets to count as inside the 'scientific community.' "[21]

With this interpretation of objectivity I have no objection,[22] although I fail to see what is provocative or particularly feminist. On the one hand, it seems simply to restate that which has been claimed all along by many traditional scientists and philosophers of science, "objectivity" as community consensus.[23] On the other hand, depending on how we count "who gets to count," "objectivity" can be interpreted as sexist, one way or another. In her most recent work, however, Harding returns to her more radical (1986) account of "objectivity," claiming that "research is socially situated, and it can be more objectively conducted without aiming for or claiming to be value-free,"[24] that "research projects [should] use their historical location as a resource for obtaining greater objectivity,"[25] and that "starting off research from women's lives will generate less partial and distorted accounts."[26]

In the final analysis, her later position is not fundamentally different from her original, provocative stance. And, like Shrader-Frechette, I cannot begin to see how starting from any moral or political bias, including the primacy of "women's lives," makes science *more* objective. Such a clearly biased starting point is, at best, irrelevant with respect to, for example, the study of quarks, and, at worst, no less insidious than the ostensive male biases already in place.

Harding must either develop this unusual account of objectivity more fully or retreat to one of the more ordinary approaches described above. Until she has done this, her account is neither worthy of criticism from traditional scientists and philosophers of science, nor deserving of defense by feminists.

Objectivity as Value-Free

The form of feminist argument against "weak objectivity no. 1"—a value-free stance—is quite simple: A value-free stance is *essential* to the scientific method; the desire to achieve a value-free stance is an *androcentric* goal; therefore, "science is a masculine project."[27] Unfortunately for the feminists, this first premise is false—a value-free stance is not essential to science or the scientific method—and hence the second premise, even if true, speaks to a straw account.

Again, the wave/particle problem, along with the Heisenberg discoveries, are cited by feminists as evidence that, even for modern science, there is no such thing as "objectivity." As I have argued above, this is a philosophical misinterpretation of empirical evidence. But even if this were not a complete misinterpretation of quantum phenomena, one can certainly recognize that "nature is no longer at arm's length"[28] and yet consistently believe that science is objective. As Stephen Toulmin has pointed out:

> We now realize, [that] the interaction between scientists and their objects of study is a two-way affair. . . . Even in fundamental physics, for instance, the fact that subatomic particles are under observation will make the influence of the physicists' instruments a significant element in the phenomena themselves . . . the scientific observer is now—willy-nilly—also a *participant*.[29]

This, though, as stated above, is more an acceptance of intersubjectivity than of subjectivity. The latter would be going too far. Toulmin has not changed what is essential to the concept of "objectivity": its commitment to unbiased evidence. He has merely restructured the traditional concept to acknowledge that we can no longer treat *any* objects (be they people or electrons) in *purely* objectified ways. This is a denial of naive scientific realism, not an acceptance of subjectivism.

Such restructuring does not depend on interpreting the concept of "objectivity" via a value-free stance; instead, it maintains the spirit of the traditional concept of "objectivity" by stressing the desire and attempt to remain unbiased. The feminist critics of objectivity seem to confuse the true belief that the scientist must realize that her position sometimes makes her both observer and observed, with the false belief that her position as scientist means that her *contextuality* cannot be overcome. She can still do good, i.e., unbiased, science.

Examples of not-quite-value-free-but-nonetheless-unbiased acts abound. They occur, for example, when we adjudicate philosophical disputes at conferences, moderate philosophical analyses in the classroom, or evaluate the work of our students. To quote Toulmin again:

> In all these cases, to be objective does not require us to be *un*interested, that is, devoid of interests or feeling; it requires us only to acknowledge those interests and feelings, to discount any resulting biases and prejudices, and to do our best to act in a *dis*interested way.[30]

Feminist criticism which is aimed at objectivity *qua* a value-free stance—value-neutrality, noninvolvedness, or disinterestedness—simply misses the point.

Objectivity as Disinterest

Some feminist critics of science and scientific methodology do in fact address the concept of "weak objectivity no. 2" in its more sophisticated form—via the notion of a disinterested or unbiased stance. Nonetheless, they still claim the traditional concept of "objectivity" is male-biased. Two different kinds of criticisms are offered.

The first focuses on a hermeneutical rendering of the texts of science as androcentric; the second focuses on the fact that "humans cannot be impartial or objective recorders of the world."[31] Both are problematic.

The Hermeneutical Fallacy*

The first criticism focuses on the fact that "objectivity" has been genderized male, while "subjectivity" has been genderized female. Both concepts are creations of an overarching "phallocentric discourse."[32]

Such generalization is obvious (to many feminists) from several different avenues: feminist historical interpretation, literary criticism, and psychoanalysis, to name a few. It is claimed that there are

---

*By "hermeneutical" I mean, simply, the interpretation of a text.

ways to " 'read science as a text' [which] reveal the social meanings—the hidden symbolic and cultural agendas—of purportedly [disinterested] claims and practices."[33] What the text has demonstrated is that science is "inextricably connected with specific masculine . . . needs and desires."[34]

This kind of "metaphor mongering"[35] is illegitimate. For one thing it has no textual support. The quotation above, for example, cites the texts of other feminist critics of science as support (e.g., see Keller), offering no specific pages, just whole texts. These texts in turn supply spotty evidence. To use Keller again as an example, when you peruse her book you find that she cites a number of other feminist critics of science who also claim to have textual evidence of male-bias in science, but has herself cited the work of only one male scientist, Francis Bacon. His use of sexist metaphor is, according to Keller, exhibited in passages like, "let us establish a chaste and lawful marriage between Mind and Nature. . . . It is Nature herself who is to be the bride, who requires taming, shaping, and subduing by the scientific mind."[36] Such passages, being so metaphorical and ambiguous could have any number of interpretations.

Moreover, no modern scientists are cited and even the outdated comments of Bacon are seen as less insidious when modified by his claim that "for man is but the servant and interpreter of nature . . . nor can nature be commanded except by being obeyed."[37] Little else is given to support this kind of claim.

But even if the list of Bacon-like metaphors was enormous, it could not do the work feminists need done. The logic is simple: such evidence presupposes precisely what is being challenged, namely, that the concept of a "disinterested stance" in traditional science is male-biased. To adopt an androcentric interpretation without offering some justification for such an adoption is to beg the question.

But what kind of justification could be offered? This is yet another problem with the account, for any appeal to evidence is problematic. Evidence is empirical and, hence, to some degree or other, the bailiwick of science. For feminists to attempt to defend their claim that there is male bias in traditional science by appealing to evidence—a concept defined within traditional science—is self-defeating.

No Such Thing as Objectivity

The hermeneutical "reason," however, is not the only justification feminist critics supply for rejecting the traditional notion of objectivity. Their other, stronger claim posits that one can never act in a disinterested way.

Why not? Why is it that "human beings cannot be impartial or objective recorders of their world?"[38] Is this a fact about the frailness of human psychology or the logical outcome of the epistemological fact that there is no disinterested stance to be had?

The Psychological Point

If feminist critics of science mean the former, then their claim, reminiscent of the psychological egoists' claim that "human beings can never act except in their own best interest," is in the same kind of trouble. As an empirical thesis it is either false, e.g., Mother Teresa, or unfalsifiable and, therefore, empty.[39]

With respect to the claim that "human beings can never act in a disinterested way," the argument against it follows suit: as an empirical thesis it is either false, e.g., when we rationally *decide,* not merely arbitrarily *choose,* which of our students ought get an "A," or unfalsifiable, hence the claim is vacuous.

James F. Harris makes an additional connection between ethical egoism and feminist criticisms of science.

> If ethical egoists really believe that people generally are selfish and ought to act in ways of furthering their own self-interests, then it seems they ought never to tell anyone about their theory since it would be against their own self-interests to have everyone else acting on an enlightened basis of such a theory. Also, an intelligent ethical egoist would not publicly announce his or her commitment to egoism since then other people would be placed on guard against the egoist. Similarly, if the feminists who advocate

> feminist science are right, if they really believe that they are right about the influence of general social and cultural values on the level of SC2 (contextual values) upon the nature of scientific inquiry on the level of SC1—[science community 1, the "object level"], then surely they ought not to publicly advocate their position in a male-dominated [science] society.[40]

To do so would find them being blackballed. It seems that feminists, like egoists, must keep their thesis, if true, to themselves.

The Epistemological Point

The psychological interpretation is probably not what feminist critics have in mind. The point is not that there are shortcomings in the human psychological mechanism which prevent one from being disinterested, but that, despite one's best intentions, there is no unbiased stance to be had. If the only stance is a biased stance then, given science and its history of male-domination, this bias translates into the fact that the male-stance is the only stance.

I am not sure if feminists who criticize science really would claim that if there is this psychological problem, that it is equally endemic to both men and women. If they say both, then feminism itself becomes one of many biases forced by one's psychological make-up. If they say that women suffer from this problem less than men, they would have to support this claim. But where could unbiased evidence for this kind of bias possibly come from?

Feminists against traditional science, however, do not directly argue for the claim that there is no unbiased stance to be had. Instead, ironically, they appeal to male authority: e.g., Thomas Kuhn. Feminists, following Kuhn, claim that the "Kuhnian strategy of arguing that observations are theory-laden, theories are paradigm-laden, and paradigms are culture-laden: [demonstrates that] there are and can be no such things as . . . objective facts."[41] Without objective facts, there can be no objective, i.e., unbiased, stance.

Of course, relying on Kuhn leaves an important question open for debate: Is he right?

Although a thorough discussion of Kuhn's arguments against the notion of "objectivity" would fall outside the scope of this [discussion], suffice it to say that, at best, there is a vast body of philosophical literature claiming he has not made his case against objectivity.[42]

Briefly, Kuhn's use of "incommensurability"—which means theories from two different paradigms cannot be compared and therefore rationally adjudicated—is at the heart of his version of relativism. Because of this, his account of relativism is caught between the horns of a dilemma. Either the thesis, on the one hand, truly embraces incommensurability or it does not. If the former, then Kuhnian relativism is provocative, for it entails unintelligibility; if the latter, it entails the promise of objectivity and is, therefore, benign. As Israel Scheffler has pointed out, "objectivity requires simply the possibility of intelligible debate over the comparative merits of rival paradigms."[43]

Although it is not clear that Kuhn himself actually supports the radical reading of the incommensurability claim, it is certain the feminists cannot simply rest on their Kuhnian laurels. Steven Yates, for example, believes that Kuhn rejects the radical reading citing Kuhn's "commensurability." He claims that Kuhn's ideas are actually quite similar to Scheffler's and that their real difference lies in the rhetoric and not the substance of their work. Yates therefore argues that "there is nothing for the feminists to exploit in any accurate reading of Kuhn; they simply do not understand him."[44]

The point is that even if Kuhn is interpreted to be a radical incommensurabilist, feminist critics of traditional science must take the body of criticisms of Kuhnian and, therefore, their own relativism seriously. If on the other hand Kuhn is not a radical incommensurabilist, then these particular feminist arguments against the traditional notion of "objectivity" cannot be Kuhn-dependent. In either case, it seems, the feminists will have to develop a Kuhnian-free attack on objectivity.

At this point it is important to mention that the feminists have not made their case against "objectivity" and, therefore, traditional scientists and philosophers of science need not reject their own projects in lieu of feminist ones. It is just as important to admit that this does not mean science—the way it is practiced by specific scientists and philosophers of science (in the laboratory, classroom, and publishing house)—is not sexist. There are certainly inequalities in these arenas. However, these are not issues that challenge what is essential to science and its methodology.

Parenthetically, some interesting examples from sociobiology, genetics, and the social sciences are given by Lynn Hankinson Nelson in her recent book *Who Knows? From Quine to a Feminist Empiricism*, see especially chapters four and five. What is noteworthy is that she admits no one has yet presented any real evidence of sexism in the hard sciences, especially physics. Nelson's feminist take on this dearth of evidence is to remain hopeful, claiming that the future will hold the answer about the sexism of physics, and we're not dead yet![45]

Furthermore, feminists must realize this lack of evidence against the (essential) male-biased nature of science is a blessing, for it is precisely by way of scientific (objective) evidence that sexism in the sciences is exposed.[46] Feminists without objectivity are feminists without evidence of sexism.

NOTES

1. Ruth Bleier, *Science and Gender: A Critique of Biology and Its theories on Women* (New York: Pergamon Press, 1984), p. 4.

2. Ibid., p. 538, n.2.

3. Evelyn Fox Keller, *Reflections on Science and Gender* (New Haven: Yale University Press, 1985), p. 12.

4. Jean Grimshaw, *Philosophy and Feminist Thinking* (Minneapolis: University of Minnesota Press, 1986), p. 83.

5. Susan Bordo, *The Flight to Objectivity* (Albany: State University of New York Press, 1987); Susan Bordo, "The Cartesian Masculinization of Thought," *Sex and Scientific Inquiry*, ed. S. Harding and J. F. O'Barr (Chicago: University of Chicago Press, 1987), pp. 247–64.

6. Nancy Tuana, review of *Sex and Scientific Inquiry*, ed. S. Harding and J. F. O'Barr, *APA Newsletter on Feminism* 89, no. 2 (1990): 62.

7. Helen Longino, "Can There Be a Feminist Science?" in *Women, Knowledge, and Reality*, ed. A. Garry and M. Pearsall (London: Unwin Hyman, 1984). p. 212.

8. Sandra Harding, *Whose Science? Whose Knowledge?* (Ithaca, N.Y.: Comell University Press, 1991), p. 143.

9. Sandra Harding, *The Science Question in Feminism* (Ithaca, N.Y.: Cornell University Press, 1986); "The Instability of the Analytical Categories of Feminist Theory," *Sex and Scientific Inquiry*, ed. S. Harding and J. F. O'Barr; Harding, *Whose Science?*; and Harding, "Starting from Women's Lives: Eight Resources for Maximizing Objectivity," *Journal of Social Philosophy* 21 (1991): 140–49.

10. Harding, *Whose Science?* p. 134.

11. Harding, *Science Question*, p. 249.

12. Harding, *Whose Science?* p. 138.

13. Ibid., p. 143.

14. Ibid.

15. Ibid., p. 144.

16. Ibid., p. 149.

17. Ibid.

18. Harding, *Science Question*, pp. 249–50, n. 7.

19. See also John Chandler, "Androcentric Science? The Science Question in Feminism," *Inquiry* 30 (1987).

20. Kristin Shrader-Frechette, review of *The Science Question in Feminism*, by Sandra Harding, *Synthese* 76 (1988): 444.

21. Harding, "Starting from Women's Lives," p. 149.

22. I have little objection for there is no reason that one cannot maintain a commitment to traditional notions of "objectivity" while walking in the large gray area between hard scientific realism and soft idealism.

23. See, for example, the works of Merleu Ponty, C. S. Peirce, Hans Reichenbach, Karl Popper, and Larry Laudan.

24. Harding, *Whose Science?* p. 159.

25. Ibid., p. 163.

26. Sandra Harding, "Rethinking Standpoint Epistemology: 'What Is Strong Objectivity?'" *Feminist Epistemologies*, ed. L. Alcoff and E. Potter, pp. 49–82.

27. Harding, "Starting from Women's Lives," p. 177, n. 5.

28. Stephen Toulmin, "The Construal of Reality: Criticism in Modern and Postmodern Science," *The Politics of Interpretation*, ed. W. J. T. Mitchell (Chicago: University of Chicago Press, 1983), p. 112.

29. Ibid., p. 103.

30. Ibid., p. 112.

31. Harding, *Science Question*, p. 83, n. 6.

32. Drucilla Cornell and Adam Thurschen, "Feminism, Negativity, Intersubjectivity," *Feminism as Critique*, ed. Seyla Benhabib and Drucilla Cornell (Minneapolis: University of Minnesota Press, 1987), p. 143.

33. Harding, *Science Question*, p. 23, n. 7.

34. Ibid., p. 23.

35. Paul R. Gross and Norman Levitt, *Higher Superstition: The Academic Left and Its Quarrels with Science* (Baltimore: Johns Hopkins University Press, 1994), p. 116.

36. Keller, *Reflections on Science*, p. 36.

37. Ibid.

38. Grimshaw, *Philosophy and Feminist Thinking*, p. 83.

39. See the criticism of psychological egoism by James Rachels, *The Elements of Moral Knowledge* (New York: McGraw-Hill, 1986) and *The Right Thing to Do* (New York: McGraw-Hill, 1989).

40. James F. Harris, *Against Relativism* (LaSalle, Ill.: Open Court, 1992), p. 184.

41. Harding, *Science Question*, p. 102.

42. Just to name a few: W. H. Newton-Smith, *The Rationality of Science* (London: Routledge and Kegan Paul, 1981); Israel Scheffler, *Science and Subjectivity* (Indianapolis: Bobbs-Merrill, 1967); Harvey Siegel, *Relativism Refitted* (Dordrecht, Holland: Reidel, 1987); Carl R. Kordig, *The Justification of Scientific Change* (Dordrecht, Holland: Reidel, 1971); and Harris, *Against Relativism.*

43. Israel Scheffler, "Vision and Revolution: A Postscript to Kuhn," *Philosophy of Science* 39 (1972): 369.

44. Steven Yates, 1992, personal telephone correspondence.

45. Lynn Hankinson Nelson, *Who Knows? From Quine to a Feminist Empiricism* (Philadelphia: Temple University Press, 1990), pp. 249–54.

46. A similar point is made by Susan Haack, "Science 'From a Feminist Perspective,' " *Philosophy* 67 (1992): 5–18.

STUDY QUESTIONS

Much of Harding's case rests on the claim that feminist scientists will have deeper, more objective insight into nature than nonfeminist scientists. In her critique, Pinnick argues that Harding has not given much evidence for this claim. For instance, feminists have not gathered a disproportionate share of Nobel Prizes. Perhaps Harding would admit that her claim is supported by little hard data, but would still argue that women's experience of oppression and marginalization, when brought to self-consciousness, gives feminists a less biased and distorted perspective. Living under oppressed circumstances might force someone to develop "street smarts"—the cognitive skills needed to survive in such conditions—but why should it enable anyone to do *science* better? On the contrary, do not oppressed people get *less* chance to develop their abilities of abstract thinking, creativity, and imagination than non-oppressed people do? Therefore, should we not expect that women's experience to be more conducive to scientific insight *after* women are no longer oppressed and marginalized?

Could conservative Christians make the same kind of argument Harding makes? Could they claim that science has adopted a dogmatic naturalism that marginalizes their views and causes creationist or intelligent design hypotheses to be rejected without a hearing? Could they not argue that the objectivity of science would be enhanced if more evangelical Christians entered the ranks of science? In short, why isn't sauce for the feminist goose also sauce for the fundamentalist gander? Presumably Harding would not be happy with this consequence and would argue that conservative Christianity is a distorting ideology but that feminism is not. Could she argue this?

Klein argues that "weak" objectivity is possible in that scientific questions can be settled by unbiased and undistorted evidence. In the more "impersonal" fields of science, such as particle physics or organic chemistry, this claim is highly plausible. Is it so plausible when we are dealing with fields where the results bear more directly on our self-image or our view of human nature? Could it be that we are not capable of achieving objectivity in the human sciences, such as psychology, sociology, and anthropology—and perhaps not in primatology either? Consider the endless debates over whether nature or nurture is more important in shaping human behavior. Can objective evidence settle these debates, or will ideology and politics always rule?

According to Harding, feminist science would be less biased and less distorting than the male-centered assumptions that currently infect science. When is a perspective less biased and less distorting than another? Is a perspective less distorting when it is closer to objective truth? But if we have criteria that show that feminism is objectively truer than competing viewpoints, why can't we have such criteria for judging a scientific claim truer than its rivals? In this case, wouldn't science be more effectively improved just by applying those criteria rather than by recruiting feminists into science? If there are no criteria for saying that feminism is objectively truer than its rivals, then is it not just one of many ideologies and just as likely to be biased as they are?

FURTHER READING

Keller, E. F. *Reflections on Gender and Science.* New Haven and London: Yale University Press, 1985.

Tuana, N., ed. *Feminism & Science.* Bloomington and Indianapolis: Indiana University Press, 1989.

PART 3

SCIENCE AND POSTMODERNISM

INTRODUCTION

In the 1980s and 1990s, "postmodernism" became a buzzword in academic circles. A buzzword is one that seems to be on everyone's lips but nobody is clear on just exactly what it means. People use buzzwords because they create an appealing "buzz" in the mind; that is, they evoke pleasing associations and confer a feeling of sophistication to the user. Add to this the fact that many of the leading postmodernist luminaries are French, and France is still a recognized trendsetter. It is easy to see why postmodernism became the latest thing in *très chic* theorizing. Though postmodernism eludes precise definition, writers classified as postmodernist wield an uncommon degree of influence, especially in departments of cultural studies, comparative literature, and other humanities fields. Because of their influence, these theorists, and especially what they say about science, need to be examined here. First though, we must get as clear as possible on what postmodernism is. The best way to get a handle is to look at the "modernism" that the more recent theory is "post."

Modernism as a worldview developed during the European Enlightenment of the eighteenth century. A salient theme of the Enlightenment was that, with modern science, a truly *objective* form of reasoning had emerged. That is, the techniques and methods of science gave us, for the first time, a way of transcending the endless metaphysical and theological squabbles that had previously occupied the best minds. Grand, but groundless, schemes of philosophical speculation, and the ever-finer logic chopping of medieval scholastics, could be set aside, and knowledge placed on a surer road. Moreso, the bitter feuds between competing systems of theology, and the fallout of inquisitions, persecutions, massacres, and religious wars, could come to an end. The motto adopted by the Royal Society, Britain's preeminent scientific body, was *"Nullius in Verba."* Loosely translated, this means "Mere words are empty"; rather one should look and see what is the case. The methods of modern science permit theories to be tested rigorously and without bias so that *nature* can settle the issue, not politics or rhetoric. Application of scientific reasoning, and banishment of ancient dogma, would make society progress toward greater knowledge, prosperity, and harmony.

The Enlightenment's dream of reason became a widespread faith in science and progress during the nineteenth century. Then, in 1914, everything came crashing down. From the Franco-Prussian War in 1870–1871 until the start of the Great War in 1914, the Western European powers had avoided armed conflict with one another (colonial wars were common, but the native peoples of Africa and Asia, lacking modern weapons, could be subdued with minor military effort). During those forty-three years, Western Europe enjoyed unprecedented technological, industrial, and scientific advancement. Social welfare movements, by improving sanitation and other living conditions, much increased the average life span. To many, it seemed that the advanced nations of Europe were starting to live the Enlightenment's dream of progress and rationality. That dream died, or was murdered rather, on Flanders Field. During four years of war, in which science was used to create new horrors of destruction, the high optimism of the Enlightenment was drowned in blood. When the slaughter finally ended, the dazed survivors, the "lost generation," could hardly be blamed for their feelings of intense disillusion and cynicism.

The postwar cynicism was especially expressed in the visual arts. Notable was Dadaism, a movement that sought to parody and debunk the supposedly "noble" ideals of art and of "high" culture in general. After all, was it not that very supposedly enlightened culture that gave the world four years of carnage? So, Dadaists produced Mona Lisas with handlebar mustaches and urinals submitted as *objets d'art*. There was a playfulness about the Dadaists, but it was a playfulness with a jagged edge. The postmodernists are the heirs of the Dadaists. Like them, the postmoderns play with "great" themes and images, freely mixing and matching the achievements of past culture but refusing to concede authority to any of them. In fact, their eclecticism is a way of asserting that no tradition is canonical and that no set of artistic or intellectual standards should have priority. So, the postmodernists also exhibit a playfulness, but, like the Dadaists, the fun has a mean streak.

Perhaps the foremost characteristic of postmodernism is this: Modernism enshrined objectivity; postmodernism celebrates relativism. Relativism, in the sense used here, is technically referred to as "epistemological relativism." Epistemology, as noted in the introduction to the previous section, is the philosophical field that seeks to define knowledge and spell out the conditions whereby knowledge claims are justified. When we seek to justify a knowledge claim, we tacitly or explicitly appeal to certain "epistemic standards"—rules or criteria that we think lead us toward truth and away from error. Sometimes epistemic standards even show up on bumper stickers, for example: "God said it. I believe it. That settles it." The person who displays this sticker is endorsing the

Bible as the standard for all knowledge claims. So, if such a person is wondering what to believe about, say, abortion or evolution, he or she will try to see what the Bible says relating to those topics and fashion his or her beliefs accordingly.

"Epistemological objectivism" may be defined as the claim that truth is absolute and so are some of the epistemic standards that guide intellectual inquiry toward such truth. That is, if a claim is true (or false), it is true (or false) independently of what humans may think, wish, or feel. Further, humans possess some standards—rules, methods, and criteria—that really are conducive to the recognition of such truth. Epistemological relativism, on the other hand, holds that truth is not absolute. Truth claims are not just so or not so, but "true" or "false" (authorized or not) relative to particular sets of epistemic standards. Since for relativists there is no absolute, capital-*T* Truth, it follows that no set of epistemic standards can lead us to such Truth. All we can say of a particular truth claim is that it is or is not sanctioned by *some* set of epistemic standards.

It follows that all truth claims are personal or parochial; they merely assert the credentials of a claim relative to the epistemic standards some person or group happens to have. For instance, the present-day astronomer views the Sun as a nuclear fusion-powered ball of mostly hydrogen and helium located approximately 93,000,000 miles from Earth. For the early Greek, the Sun was the blazing chariot of Helios, the sun god, whose course across the sky was just barely above the mountaintops. For the relativist, neither view of the Sun is truer in any absolute sense than the other; each is just a reflection of the standards guiding knowledge at very different times and places. Therefore, there is no absolute scale for ranking some set of epistemic criteria as better than another. If there is no absolute Truth "out there," it is obvious that no set of epistemic standards can be objectively better at leading us to such mythical Truth.

Well, what is wrong with such relativism? It is an attractive option for many people. Relativism seems to imply that, since no person or group can claim access to absolute truth, we should listen tolerantly and patiently to all the voices in the "conversation of mankind." Hopi wisdom should be heard just as seriously as particle physics since Hopis and physicists are each just telling their own stories. Again, such views are appealing to many people. When actress-turned-New-Age-guru Shirley MacLaine was asked about the basis of her belief in reincarnation, she replied that reincarnation was "true for her." She seemed to be endorsing relativism by claiming that reincarnation is "true" or "false" relative to a personal set of epistemic standards.

Despite its admitted attractions, such as apparently promoting a liberal and tolerant view of other peoples, customs, and cultures, relativism is vigorously rejected by most philosophers and nearly all scientists. Philosopher John Searle points out that, like computers, the human mind comes with default settings. That is, barring compelling reason to think otherwise, we just naturally and spontaneously take some things for granted. For instance, when we have the experience of seeing a certain very thick book misplaced on the bedroom dresser, we spontaneously and naturally think that the telephone directory *really is* on the dresser. Also, we think that seeing it *really is* an excellent reason for thinking that it is on the dresser. If someone asks us where the telephone directory is, we answer, "I just saw it on the bedroom dresser." Our answer indicates that we are very confident about where the item is and our grounds for saying that it is really there. Innumerable such homely examples indicate that the default setting for humans is to regard some claims as simply (absolutely) true and some of our epistemic standards ("seeing is believing") as reliable guides to such truth. This spontaneous epistemological objectivism extends to such claims as that the Sun averages 93,000,000 miles distance from Earth or even that God exists. People who assert these things nearly always are claiming that they are *really* so, not just "true for them," and further that there are good reasons for holding these claims so.

Now default settings on computers or humans can be changed, but we humans have the right to ask for *very* good reasons for changing them. When I hear characteristic scratchings and mewings, I form the opinion "The cat is at the door." When I spontaneously and unreflectively form this opinion, I think the cat *really is* at the door. I rightly demand very good reason to instead regard "The cat is at the door" as merely a belief sanctioned by culturally local and historically arbitrary standards. The same feeling applies to many scientific or historical claims. For instance, if someone were to tell me that there are no objective reasons for preferring the claims of legitimate scientists or historians over those of creationists or Holocaust deniers, I would expect some *very* good arguments. So, if objectivism is our null or default hypothesis, have relativists given us good reason to abandon it? All that I can say here is that in many years of following relativist arguments, I have yet to find such reasons.

One obvious problem with relativism is that it is very hard to state in a way that is not self-defeating. Consider the claim "Epistemological relativism is true." Let us call this claim *R*. How should we take *R*, as absolutely true or only as relatively true? If relativists say that *R* is absolutely true, then they abandon relativism since they now admit that there is one absolute truth, namely *R*. Worse, "Epistemological relativism is (absolutely) true" is just incoherent; if *R* is absolutely true, then there is no absolute truth, so *R* cannot be absolutely true. If the claim is that *R* is relatively true, that is, that it is "true" relative to the epistemic standards accepted by relativists, then objectivists will rightly react to this unsurprising revelation with a shrug. Of course relativism is "true" for relativists! Over the centuries, relativists have struggled mightily to wiggle out of this dilemma—unsuccessfully in my view.

For most postmodernists, relativism is not so much an explicit doctrine as a *modus operandi;* it is a style more than a stance. That is, their attitude toward any traditional standard or canon is inevitably flippant, ironic, or skeptical. This attitude extends even to the accepted canons of clear and precise writing. Their prose is often verbose, jargon-laden, filled with puns, and deliberately ambiguous (thus illustrating their rejection of the canonical view that language should have fixed, definite meanings). However obscure their rhetoric, though, the postmodernists' attitude toward science is clear. They want to debunk science, to remove it from its pedestal and cut it down to size. The want to show that science is just one "form of discourse" among indefinitely many, and no more "rational" or "objective" than any other.

The first reading in this section is from Donna Haraway's *Primate Visions: Gender, Race, and Nature in the World of Modern Science.* Haraway could just as easily have been included in the section on feminism. Postmodernist feminism is one very influential branch of feminism, and much postmodernist thought focuses on feminist issues. The introduction to *Primate Visions,* which is included here, outlines her aims and assumptions in the book. She is writing about primates because they are "popular, important, marvelously varied, and controversial." Yet, she makes it clear that she does not intend to write a disinterested, objective study, which, she claims, is impossible anyway. Science purports to present facts, but Haraway thinks that science is best seen as an evolving craft for telling stories. Indeed, she does not accept the traditional dichotomy between fact and fiction. Rather, she sees scientific "fact" as merely a particular kind of narrative, one constrained by the rules and conventions of scientific practice. Her intention therefore is to regard the narratives of primatology as a kind of science fiction.

The overall aim of *Primate Visions* is to offer a subversive reading of primatology, one that undermines what Haraway calls "binarisms." A binarism is an assumed conceptual dualism, one that divides things into two opposed and mutually exclusive categories, such as fact vs. fiction, nature vs. nurture, subject vs. object, or sex vs. gender. Much postmodernist writing is an attempt to undermine or "deconstruct" such dichotomies. Haraway sees primatology, as it has been practiced, as a set of narratives that have been used to justify and extend certain of these oppressive binarisms. She holds that primates have been portrayed as the primal "other," the "them" who stand opposed to "us," and who represent a "natural" state that is prior to human culture. Yet if, as Haraway believes, the "nature" of primates is itself a set of biased, culturally constructed concepts, those concepts can be used to justify as "natural" all sorts of oppressive stereotypes. Recall what was said in the introduction to the previous section about reading the "naturalness" of the primate alpha male's dominance back into human society. Haraway thinks that by exposing the underlying political, ideological, and perhaps racist and sexist, agendas of the narratives of primatology, she will undermine the oppressive dualisms that have unfairly stereotyped and marginalized people.

More traditional scholars often regard postmodernism as pedantic nonsense, if not an outright fraud. Anyone sympathetic with this view will regard the next selection, Matt Cartmill's review of *Primate Visions,* as delightfully malicious. Cartmill pulls no punches, but his review is not a mere diatribe; he raises some very serious questions about Haraway's approach and conclusions. Cartmill charges that, as one might expect given her denigration of objectivity, Haraway often presents skewed evidence and caricatures of persons and positions she dislikes. Indeed, he claims that her whole book is a farrago of irrelevance, half-truths, and inconsistency, all expressed in a turgid and pretentious style. Much of her case, says Cartmill, rests upon far-fetched connections that she claims to see between things that are unrelated or only coincidentally associated.

In the introduction I mentioned the Sokal hoax as the focal point of much of the dispute about postmodernist and constructivist views of science. Briefly, physicist Alan Sokal wrote a parody of postmodernist/constructivist science criticism and sent it, as an apparently serious submission, to the periodical *Social Text. Social Text* took Sokal's essay seriously and published it. Unfortunately, Sokal's essay, "Transgressing the Boundaries," is too long and technical to reproduce here. However, we include an essay from the *New York Review of Books* in which physicist Steven Weinberg, a winner of the Nobel Prize, reflects on the hoax and its implications.

Weinberg thinks that Sokal revealed the scientific ignorance of the postmodernist and constructivist science critics. For instance, he notes that many such critics focus on the terms "linear" and "nonlinear," which have precise mathematical meanings. However, postmodernist intellectuals, stimulated by the metaphorical resonance of these terms, but clueless about their meanings, have turned them into buzzwords where "nonlinear" signifies the up-to-date and "linear" stigmatizes the retrograde. Weinberg notes that one science critic referred to quantum theory as "nonlinear" when in fact it is the only known precisely linear theory. Another misuses the term "unified field theory" and another alleges that mathematicians have neglected spaces with boundaries when there is actually a large literature on the subject. In case after case, science is criticized by those who do not even understand scientific terminology. Weinberg also criticizes postmodernists, and a number of scientists, for trying to read large philosophical or cultural consequences into the discoveries of science. He claims that there are no such extrascientific messages. Finally, Weinberg criticizes postmodernists for denying the objective reality of the laws of physics. Such laws are not like the rules of baseball, as some postmodernists assert; they are no more our invention than rocks, says Weinberg.

Several critics responded to Weinberg, saying that in criticizing the postmodernists he had gone too far in the opposite direction, creating an ideology of science as wholly separate from culture. Michael Holquist and Robert Schulman, while distancing themselves from the editors of *Social Text,* charge that Weinberg has introduced a dualism between science and culture reminiscent of the religious distinction between the sacred and the profane. Further, they charge that Weinberg, a particle physicist, has set up particle physics as the model of "true" science and appointed himself as the guardian of scientific purity. Similarly, historian of physics M. Norton Wise charges that Weinberg has unfairly stigmatized historians, philosophers, and sociologists of science as deniers of objectivity and as uninterested in rational understanding. Conversely, says Wise, many scientists have been mystics, and the discoverers of, for instance, the laws of electromagnetism—Oersted, Faraday, Lord Kelvin, and Maxwell—were deeply religious. Wise charges that Weinberg's polarization of science and culture would ban such mystical physicists from the field.

Weinberg replies to Holquist and Schulman that his view is indeed dualist; he thinks it is essential to recognize a gap between science and human culture. For instance, contrary to the many claims about what follows from quantum mechanics (e.g., mysticism, free will), none of those things is implied by the mathematical formalism of the theory. True, quantum theory, which is indeed eerie, may inspire or provide metaphors, but there is no logical entailment of such alleged consequences. To Professor Wise he replies that he never denied that scientists were often inspired by culture and has no wish to drive mystics from the field. His contention is that, whatever motivation or insight physicists may have found in such factors, the social and mystical elements did not become permanent features of their theories. For instance, everyone who comprehends electricity and magnetism now understands Maxwell's equations in the same way; Maxwell's cultural context and religious convictions are now irrelevant.

EIGHT

INTRODUCTION TO *PRIMATE VISIONS*

DONNA HARAWAY

The names you uncaged primates give things affect your attitude to them forever after.

–Ruth Herschberger, *Adam's Rib* (New York: Harper and Row, 1948)

For thus all things must begin, with an act of love.

–Eugene Marais, "Soul of the White Ant," Africa Radio Broadcasting, 1980

How are love, power, and science intertwined in the constructions of nature in the late twentieth century?[1] What may count as nature for late industrial people? What forms does love of nature take in particular historical contexts? For whom and at what cost? In what specific places, out of which social and intellectual histories, and with what tools is nature constructed as an object of erotic and intellectual desire? How do the terrible marks of gender and race enable and constrain love and knowledge in particular cultural traditions, including the modern natural sciences? Who may contest for what the body of nature will be? These questions guide my history of the modern sciences and popular cultures emerging from accounts of the bodies and lives of monkeys and apes.

The themes of race, sexuality, gender, nation, family, and class have been written into the body of nature in Western life sciences since the eighteenth century. In the wake of post–World War II decolonization, local and global feminist and antiracist movements, nuclear and environmental threats, and broad consciousness of the fragility of the earth's webs of life, nature remains a crucially important and deeply contested myth and reality. How do material and symbolic threads interweave in the fabric of late-twentieth-century nature for industrial people?

Monkeys and apes have a privileged relation to nature and culture for Western people: simians occupy the border zones between those potent mythic poles. In the border zones, love and knowledge are richly ambiguous and productive of meanings in which many people have a stake. The commercial and scientific traffic in monkeys and apes is a traffic in meanings, as well as in animal lives. The sciences that tie monkeys, apes, and people together in a Primate Order are built through disciplined practices deeply enmeshed in narrative, politics, myth, economics, and technical possibilities. The women and men who have contributed to primate studies have carried with them the marks of their own histories and cultures. These marks are written into the texts of the lives of monkeys and apes, but often in subtle and unexpected ways. People who study other primates are advocates of contending scientific discourses, and they are accountable to many kinds of audiences and patrons. These people have engaged in dynamic, disciplined, and intimate relations of love and knowledge with the animals they were privileged to watch. Both the primatologists and the animals on whose lives they reported command intense popular interest—in natural history museums, television specials, zoos, hunting, photography, science fiction, conservation politics, advertising, cinema, science news, greeting cards, and jokes. The animals have been claimed as privileged subjects by disparate life and human sciences—anthropology, medicine, psychiatry, psychobiology, reproductive physiology, linguistics, neural biology, paleontology, and behavioral ecology. Monkeys and apes have modeled a vast array of human problems and hopes. Most of all, in European, American, and Japanese societies, monkeys and apes have been subjected to sustained, culturally specific interrogations of what it means to be "almost human."

Monkeys and apes—and the people who construct scientific and popular knowledge about them—are part of cultures in contention. Never innocent, the visualizing narrative "technology" of this book draws from contemporary theories of cultural production, historical and social studies of science and technology, and feminist and antiracist movements and theories to craft a view of nature as it is constructed and reconstructed in the bodies and lives of "third world" animals serving as surrogates for "man."

I have tried to fill *Primate Visions* with potent verbal and visual images—the corpse of a gorilla shot in 1921 in the "heart of Africa" and transfixed into a lesson in civic virtue in the American Museum of Natural History in New York City; a little white girl brought into the Belgian Congo in the 1920s to hunt gorilla with a camera, who metamorphosized in the 1970s into a writer of science fiction considered for years as a model of masculine prose; the chimpanzee HAM in his space capsule in the Mercury Project in 1961; HAM's chimp contemporary, David Greybeard, reaching out to Jane Goodall, "alone" in the "wilds of Tanzania" in the year in which fifteen African primate-habitat nations achieved national independence; a *Vanity Fair* special on the murdered Dian Fossey in a gorilla graveyard in Rwanda in 1986; the bones of an ancient fossil, reconstructed as the grandmother of humanity, laid out like jewels on red velvet in a paleontologist's laboratory in a pattern to ground, once again, a theory of the origin of "monogamy"; infant monkeys in Harry Harlow's laboratory in the 1960s clinging to cloth and wire "surrogate mothers" at a historical moment when the images of surrogacy began to surface in American reproductive politics; the emotionally wrenching embrace between a young, middle-class, white woman scientist and an adult American Sign Language–speaking chimpanzee on an island in the River Gambia, where white women teach captive apes to "return" to the "wild"; a Hallmark greeting card reversing the images of King Kong with a monstrous blond woman and a cringing silverback gorilla in bed in a drama called "Getting Even"; the anatomical drawings of living and fossil female apes sharing the basic lines of their bodies with a modern human female, in order to teach medical students the functional meaning of human adaptations; ordinary women and men from Africa, the United States, Japan, Europe, India, and elsewhere, with tape recorders and data clipboards transcribing the lives of monkeys and apes into specialized texts that become contested items in political controversies in many cultures.

I am writing about primates because they are popular, important, marvelously varied, and controversial. And all members of the Primate Order—monkeys, apes, and people—are threatened. Late-twentieth-century primatology may be seen as part of a complex survival literature in global, nuclear culture. Many people, including myself, have emotional, political, and professional stakes in the production and stabilization of knowledge about the order of primates. This will not be a disinterested, objective study, nor a comprehensive one—partly because such studies are impossible for anyone, partly because I have stakes I want to make visible (and probably others as well). I want this book to be interesting for many audiences, and pleasurable and disturbing for all of us. In particular, I want this book to be responsible to primatologists, to historians of science, to cultural theorists, to the broad left, antiracist, anticolonial, and women's movements, to animals, and to lovers of serious stories. It is perhaps not always possible to be accountable to those contending audiences, but they have all made this book possible. They are all inside this text. Primates existing at the boundaries of so many hopes and interests are wonderful subjects with whom to explore the permeability of walls, the reconstitution of boundaries, the distaste for endless socially enforced dualisms.

FACT AND FICTION

Both science and popular culture are intricately woven of fact and fiction. It seems natural, even morally obligatory, to oppose fact and fiction; but their similarities run deep in Western culture and language. Facts can be imagined as original, irreducible nodes from which a reliable understanding of the world can be constructed. Facts ought to be discovered, not made or constructed. But the etymology of facts refers us to human action, performance, indeed, to human feats (OED). Deeds, as opposed to words, are the parents of facts. That is, human action is at the root of what we can see as a fact, linguistically and historically. A fact is the thing done, a neuter past participle in our Roman parent language. In that original sense, facts are what has actually happened. Such things are known by direct experience, by testimony, and by interrogation—extraordinarily privileged routes to knowledge in North America.

Fiction can be imagined as a derivative, fabricated version of the world and experience, as a kind of perverse double for the facts or as an escape through fantasy into a better world than "that which actually happened." But tones of meaning in fiction make us hear its origin in vision, inspiration, insight, genius. We hear the root of fiction in poetry and we believe, in our Romantic moments, that original natures are revealed in good fiction. That is, fiction can be *true*, known to be true by an appeal to nature. And as nature is prolific, the mother of life in our major myth systems, fiction seems to be an inner truth which gives birth to our actual lives. This, too, is a very privileged route to knowledge in Western cultures, including the United States. And finally, the etymology of fiction refers us once again to human action, to the act of fashioning, forming, or inventing, as well as to feigning. Fiction is inescapably implicated in a dialectic of the true (natural) and the counterfeit (artifactual). But in all its meanings, fiction is about human action. So, too, are all the narratives of science—fiction and fact—about human action.

Fiction's kinship to facts is close, but they are not identical twins. Facts are opposed to opinion, to prejudice, but not to fiction. Both fiction and fact are rooted in an epistemology that appeals to experience. However, there is an important difference; the word *fiction* is an active form, referring to a present act of fashioning, while *fact* is a descendant of a past participle, a word form which masks the generative deed or performance. A fact seems done, unchangeable, fit only to be recorded; fiction seems always inventive, open to other possibilities, other fashionings of life. But in this opening lies the threat of merely feigning, of not telling the true form of things.

From some points of view, the natural sciences seem to be crafts for distinguishing between fact and fiction, for substituting the past participle for the invention, and thus preserving true experience from its counterfeit. For example, the history of primatology has been repeatedly told as a progressive clarification of sightings of monkeys, apes, and human beings. First came the original intimations of primate form, suggested in the prescientific mists in the inventive stories of hunters, travelers, and natives, beginning perhaps in ancient times, perhaps in the equally mythic Age of Discovery and of the Birth of Modern Science in the sixteenth century. Then gradually came clear-sighted vision, based on anatomical dissection and comparison. The story of correct vision of primate social form has the same plot: progress from misty sight, prone to invention, to sharp-eyed quantitative knowledge rooted in that kind of experience called, in English, experiment. It is a story of progress from immature sciences based on mere description and free qualitative interpretation to mature science based on quantitative methods and falsifiable hypotheses, leading to a synthetic scientific reconstruction of primate reality. But these histories are stories about stories, narratives with a good ending; i.e., the facts put together, reality reconstructed scientifically. These are stories with a particular aesthetic, realism, and a particular politics, commitment to progress.

From only a slightly different perspective, the history of science appears as a narrative about the history of technical and social means to produce the facts. The facts themselves are types of stories, of testimony to experience. But the provocation of experience requires an elaborate technology—including physical tools, an accessible tradition of interpretation, and specific social relations. Not just anything can emerge as a fact; not just anything can be seen or done, and so told. Scientific practice may be considered a kind of story-telling practice—a rule-governed, constrained, historically changing craft of narrating the history of nature. Scientific practice and scientific theories produce and are embedded in particular kinds of stories. Any scientific statement about the world depends intimately upon language, upon metaphor. The metaphors may be mathematical or they may be culinary; in any case, they structure scientific vision. Scientific practice is above all a story-telling practice in the sense of historically specific practices of interpretation and testimony.

Looking at primatology, a branch of the life sciences, as a story-telling craft may be particularly appropriate. First, the discourse of biology, beginning near the first decades of the nineteenth century, has been about organisms, beings with a life history; i.e., a plot with structure and function.[2] Biology is inherently historical, and its form of discourse is inherently narrative. Biology as a way of knowing the world is kin to Romantic literature, with its discourse about organic form and function. Biology is the fiction appropriate to objects called organisms; biology fashions the facts "discovered" from organic beings. Organisms perform for the biologist, who transforms that performance into a truth attested by disciplined experience; i.e., into a fact, the jointly accomplished deed or feat of the scientist and the organism. Romanticism passes into realism, and realism into naturalism, genius into progress, insight into fact. *Both* the scientist and the organism are actors in a story-telling practice.

Second, monkeys, apes, and human beings emerge in primatology inside elaborate narratives about origins, natures, and possibilities. Primatology is about the life history of a taxonomic order that includes people. Especially Western people produce stories about primates while simultaneously telling stories about the relations of nature and culture, animal and human, body and mind, origin and future. Indeed, from the start, in the mid-eighteenth century, the primate order has been built on tales about these dualisms and their scientific resolution.

To treat a science as narrative is not to be dismissive, quite the contrary. But neither is it to be mystified and worshipful in the face of a past participle. I am interested in the narratives of scientific fact—those potent fictions of science—within a complex field indicated by the signifier SF. In the late 1960s science fiction anthologist and critic Judith Merril idiosyncratically began using the signifier SF to designate a complex emerging narrative field in which boundaries between science fiction (conventionally, SF) and fantasy became highly permeable in confusing ways, commercially and linguistically. Her designation, SF, came to be widely adopted as critics, readers, writers, fans, and publishers struggled to comprehend an increasingly heterodox array of writing, reading, and marketing practices indicated by a proliferation of "SF " phrases: speculative fiction, science fiction, science fantasy, speculative futures, speculative fabulation.

SF is a territory of contested cultural reproduction in high-technology worlds. Placing the narratives of scientific fact within the heterogeneous space of SF produces a transformed field. The transformed field sets up resonances among all of its regions and components. No region or component is "reduced" to any other, but reading and writing practices respond to each other across a structured space. Speculative fiction has different tensions when its field also contains the inscription practices that constitute scientific fact. The sciences have complex histories in the constitution of imaginative worlds and of actual bodies in modern and postmodern "first world" cultures. Teresa de Lauretis speculated that the sign work of SF was "potentially creative of new forms of social imagination, creative in the sense of mapping out areas where cultural change *could* take place, of envisioning a different order of relationships between people and between people and things, a different conceptualization of social existence, inclusive of physical and material existence."[3] This is also one task of the "sign work" of primatology.

So, in part, *Primate Visions* reads the primate text as science fiction, where possible worlds are constantly reinvented in the contest for very real, present worlds. The conclusion perversely reads a SF story about an alien species that intervenes in human reproductive politics as if it were a monograph from the primate field. Beginning with the myths, sciences, and historical social practices that placed apes in Eden and apes in space, at the beginnings and ends of Western culture, *Primate Visions* locates aliens in the text as a way to understand love and knowledge among primates on a contemporary fragile Earth.

FOUR TEMPTATIONS

Analyzing a scientific discourse, primatology, as story-telling within several contested narrative fields is a way to enter current debates about the social construction of scientific knowledge without succumbing completely to any of four very tempting positions, which are also major resources for the approaches [taken here]. I use the image of temptation because I find all four positions persuasive, enabling, and also dangerous, especially if any one position finally silences all the others, creating a false harmony in the primate story.

The first resourceful temptation comes from the most active tendencies in the social studies of science and technology. For example, the French prominent analyst of science, Bruno Latour, radically rejects all forms of epistemological realism and analyzes scientific practice as thoroughly social and constructionist. He rejects the distinction between social and technical and represents scientific practice as the refinement of "inscription devices," i.e., devices for transcribing the immense complexity and chaos of competing interpretations into unambiguous traces, writings, which mark the emergence of a fact, the case about reality. Interested in science as a fresh form of power in the social-material world and scientists as investing "their political ability in the heart of doing science," Latour and his colleague Stephen Woolgar powerfully describe how processes of construction are made to invert and appear in the form of discovery.[4] The accounts of the scientists about their own processes become ethnographic data, subject to cultural analysis.

Fundamentally, from the perspective of *Laboratory Life*, scientific practice is literary practice, writing, based on jockeying for the power to stabilize definitions and standards for claiming something to be the case. To win is to make the cost of destabilizing a given account too high. This approach can explain scientific contests for the power to close off debate, and it can account for both successful and unsuccessful entries in the contest. Scientific practice is negotiation, strategic moves, inscription, translation. A great deal can be said about science as effective belief and the world-changing power to enforce and embody it.[5] What more can one ask of a theory of scientific practice?

The second valuable temptation comes from one branch of the Marxist tradition, which argues for the historical superiority of particular structured standpoints for knowing the social world, and possibly the "natural" world as well. Fundamentally, people in this tradition find the social world to be structured by the social relations of the production and reproduction of daily life, such that it is only possible to see these relations clearly from some vantage points. This is not an individual matter, and good will is not at issue. From the standpoint of those social groups in positions of systematic domination and power, the true nature of social life will be opaque; they have too much to lose from clarity.

Thus, the owners of the means of production will see equality in a system of exchange, where the standpoint of the working class will reveal the nature of domination in the system of production based on the wage contract and the exploitation and deformation of human labor. Those whose social definition of identity is rooted in the system of racism will not be able to see that the definition of human has not been neutral, and cannot be until major material-social changes occur on a world scale. Similarly, for those whose possibility of adult status rests on the power to appropriate the "other" in a sociosexual system of gender, sexism will not look like a fundamental barrier to correct knowledge *in general.* The tradition indebted to Marxist epistemology can account for the greater adequacy of some ways of knowing and can show that race, sex, and class fundamentally determine the most intimate details of knowledge and practice, especially where the appearance is of neutrality and universality.[6]

These issues are hardly irrelevant to primatology, a science practiced in the United States nearly exclusively by white people, and until quite recently by white men, and still practiced overwhelmingly by the economically privileged. Much of [my] book [on this subject] examines the consequences for primatology of the social relations of race, sex, and class in the construction of scientific knowledge. For example, perhaps most primatologists in the field in the first decades after World War II failed to appreciate that the interrelationships of people, land, and animals in Africa and Asia are at least partly due to the positions of the researchers within systems of racism and imperialism. Many sought "pure" nature, unspoiled by contact with people; and so they sought untouched species, analogous to the "natives" once sought by colonial anthropologists. But for the observer of animals, the indigenous peoples of Africa and Asia were a nuisance, a threat to conservation—indeed, encroaching "aliens"—until decolonization forced white Western scientists to restructure their biopolitics of self and other, native and alien. The boundaries among animals and human beings shift in the transition from a colonial to a post- or neocolonial standpoint. Insisting that there can be less deformed contents and methods in the natural as well as social sciences, the Marxist, feminist, and antiracist accounts reject the relativism of the social studies of science. Explicitly political accounts take sides on what is a more adequate, humanly acceptable knowledge. But these analyses have limits for guiding an exploration of primate studies. Wage labor, sexual and reproductive appropriation, and racial hegemony are structured aspects of the human social world. There is no doubt that they affect knowledge systematically, but it is not clear precisely how they relate to knowledge about the feeding patterns of patas monkeys or about the replication of DNA molecules.

Another aspect of the Marxist tradition has made significant progress in answering that kind of question. In the 1970s, people associated with the British *Radical Science Journal* developed the concept of science as a labor process in order to study and change scientific mediations of class domination in the relations of production and reproduction of human life.[7] Like Latour, they leave no holes for a realist or positivist epistemology, the preferred versions of most practicing scientists. Every aspect of scientific practice can be described in terms of the concept of mediation: language, laboratory hierarchies, industrial ties, medical doctrines, basic theoretical preferences, and stories about nature. The concept of labor process seems cannibalistic, making the social relations of other basic processes seem derivative. For example, the complex systems of domination, complicity, resistance, equality, and nurturance in gendered practices of bearing and raising children cannot be accommodated by the concept of labor. But these reproductive practices visibly affect more than a few contents and methods in modern primate studies. But even an extended concept of mediation and systematic social process, one that does not insist on the reduction to labor in a classic Marxist sense, leaves out too much.

The third temptation comes from the siren call of the scientists themselves; they keep pointing out that they are, among other things, watching monkeys and apes. In some sense, more or less nuanced, they insist that scientific practice "gets at" the world. They claim that scientific knowledge is not simply about power and control. They claim that their knowledge somehow translates the active voice of their subjects, the objects of knowledge. Without necessarily being compelled by their aesthetic of realism or their theories of representation, I believe them in the strong sense that my imaginative and intellectual life and my professional and political commitments in the world respond to these scientific accounts. Scientists are adept at providing good grounds for belief in their accounts and for action on their basis. Just how science "gets at" the world remains far from resolved. What does seem resolved, however, is that science grows from and enables concrete ways of life, including particular constructions of love, knowledge, and power. That is the core of its instrumentalism and the limit to its universalism.

Evidence is always a question of interpretation; theories are accounts *of* and *for* specific kinds of lives. I am looking for a way of telling a story of the production of a branch of the life sciences, a branch which includes human beings centrally, that listens very carefully to the stories themselves. My story must listen to the practices of interpretation of the primate order in which the primates themselves—monkeys, apes, and people—all have some kind of "authorship." I would suggest that the concept of constrained and contested story-telling allows an appreciation of the social construction of science, while still guiding the hearer to a search for the other animals who are active participants in primatology. I want to find a concept for telling a history of science that does not itself depend on the dualism between active and passive, culture and nature, human and animal, social and natural.

The fourth temptation intersects with each of the other three; this master temptation is to look always through the lenses ground on the stones of the complex histories of gender and race in the constructions of modern sciences around the globe. That means examining cultural productions, including the primate sciences, from the points of view enabled by the politics and theories of feminism and antiracism. The challenge is to remember the particularity as well as the power of this way of reading and writing. But that is the same challenge that should be built into reading or writing a scientific text. Race and gender are not prior universal social categories—much less natural or biological givens. Race and gender are the world-changing products of specific, but very large and durable, histories. The same thing is true of science. The visual system of [my argument] depends upon a triple filter of race, gender, and science. This is the filter which traps the marked bodies of history for closer examination.[8]

Stories are always a complex production with many tellers and hearers, not all of them visible or audible. Story-telling is a serious concept, but one happily without the power to claim unique or closed readings. Primatology seems to be a science composed of stories, and the purpose of this [essay] is to enter into contestations for their construction. The lens of story-telling defines a thin line between realism and nominalism; but primates seem to be natural scholastics, given to equivocation when pressed. Also, I think there is an aesthetic and an ethic built into thinking of scientific practice as story-telling, an aesthetic and ethic different from capitulation to "progress" and belief in knowledge as passive reflection of "the way things are," and also different from the ironic skepticism and fascination with power so common in the social studies of science. The aesthetic and ethic latent in the examination of story-telling might be pleasure and responsibility in the weaving of tales. Stories are means to ways of living. Stories are technologies for primate embodiment.

PRIMATOLOGY IS (JUDEO-)CHRISTIAN SCIENCE

Western Jews and Christians or post-Judeo-Christians are not the only practitioners of primate sciences. But I focus primarily on the history of studies of the social behavior of monkeys and apes done in the United States or by Euro-Americans in the twentieth century. In these stories, there is a constant refrain drawn from salvation history; primatology is about primal stories, the origin and nature of "man," and about reformation stories, the reform and reconstruction of human nature. Implicitly and explicitly, the story of the Garden of Eden emerges in the sciences of monkeys and apes, along with versions of the origin of society, marriage, and language. From the beginning, primatology has had this character in the West. If the eighteenth-century Swedish "father" of modern biological classification, Linnaeus, is cited at all by twentieth-century scientists, he is noted for placing human beings in a taxonomic order of nature with other animals, i.e., for taking a large step away from Christian assumptions. Linnaeus placed "man" in his taxonomic order of Primates as *Homo sapiens,* in the same genus with *Homo troglodytes,* a dubious and interesting creature illustrated as a hairy woman in Linnaeus's probable source. Also in the new primate order in the tenth edition of the *Systema Naturae* of 1758 were a genus for monkeys and apes, one for lemurs, and one for bats. But there is quite another way to see Linnaeus's activity as the "father" of a discourse about nature. He referred to himself as a second Adam, the "eye" of God, who could give true representations, true names, thus reforming or restoring a purity of names lost by the first Adam's sin. Nature was a theatre, a stage for the playing out of natural and salvation history. The role of the one who renamed the animals was to ensure a true and faithful order of nature, to purify the eye and the word. The "balance of nature" was maintained partly by the role of a new "man" who would see clearly and name accurately, hardly a trivial identity in the face of eighteenth-century European expansion. Indeed, this is the identity of the modern authorial subject, for whom inscribing the body of nature gives assurance of his mastery.

Linnaeus's science of natural history was intimately a Christian science. Its first task, achieved in Linnaeus's and his correspondents' life work, was to announce the kinship of "man" and beast in the modern order of an expanding Europe. Natural man was found not only among the "savages," but also among the animals, who were named primates in consequence, the first order of nature. Those who could bestow such names had a powerful modern vocation; they became scientists. Taxonomy had a secular sacred function. The "calling" to practice science has kept this sacralized character into the late twentieth century, although we will see it at its strongest in the early part of our century. The stories produced by such practitioners have a special status in a repressed Protestant biblical culture like that of the United States.

Nature for Linnaeus was not understood "biologically," but "representationally." In the course of the nineteenth century, biology became a discourse about productive, expanding nature. Biology was constructed as a discourse about nature known as a system of production and reproduction, understood in terms of the functional division of labor and the mental, labor, and sexual efficiency of organisms. In this context, by the twentieth century primates were cast into an *Ecological Theatre and an Evolutionary Play*.[9] The drama has been about the origin and development of many persistent mythic themes: sex, language, authority, society, competition, domination, cooperation, family, state, subsistence, technology, and mobility. There are two major readings of the play adopted in [my] book: One attends to symbolic meanings, to the primate sciences as a kind of art form making repeated use of the narrative resources of Judeo-Christian myth systems. The second pays particular attention to the ways primate biology is theorized as a material system of production and reproduction, a kind of "materialist" reading. Both interpretations listen for echoes and determinants of race, sex, and class in the stories. The primate body, as part of the body of nature, may be read as a map of power. Biology, and primatology, are inherently political discourses, whose chief objects of knowledge, such as organisms and ecosystems, are icons (condensations) of the whole of the history and politics of the culture that constructed them for contemplation and manipulation. The primate body itself is an intriguing kind of political discourse.

PRIMATOLOGY IS SIMIAN ORIENTALISM

The argument [here] is that primatology is about an Order, a taxonomic and *therefore* political order that works by the negotiation of boundaries achieved through ordering differences. These boundaries mark off important social territories, like the norm for a proper family, and are established by social practice, like curriculum development, mental health policy, conservation politics, film making, and book publishing. The two major axes structuring the potent scientific stories of primatology that are elaborated in these practices are defined by the interacting dualisms, *sex/gender* and *nature/culture*. Sex and the West are axiomatic in biology and anthropology. Under the guiding logic of these complex dualisms, Western primatology is simian orientalism.

Edward Said argued that Western (European and American) scholars have had a long history of coming to terms with countries, peoples, and cultures in the Near and Far East that is based on the Orient's special place in Western history—the scene of origins of language and civilization, of rich markets and colonial possession and penetration, and of imaginative projection.[10] The Orient has been a troubling resource for the production of the Occident, the "East's" other and periphery that became materially its dominant. The West is positioned outside the Orient, and this exteriority is part of the Occident's practice of representation. Said quotes Marx, "They cannot represent themselves; they must be represented."[11] These representations are complex mirrors for Western selves in specific historical moments. The West has also been positioned mobilely; Westerners could be *there* with relatively little resistance from the other. The difference has been one of power. The structure has been limiting, of course, but more importantly, it has been *productive.* That productivity occurred within the structured practices and discourses of orientalism; the structures were a condition of having anything to say. There never is any question of having anything truly original to say about origins. Part of the authority of the practices of telling origin stories resides precisely in their intertextual relations.

Without stretching the comparison too far, the signs of orientalist discourse mark primatology. But here, the scene of origins is not the cradle of civilization, but the cradle of culture, of human being distinct from animal existence. If orientalism concerns the Western imagination of the origin of the city, primatology displays the Western imagination of the origin of sociality itself, especially in the densely meaning-laden icon of "the family." Origins are in principle inaccessible to direct testimony; any voice from the time of origins is structurally the voice of the other who generates the self. That is why both realist and postmodernist aesthetics in primate representations and simulations have been modes of production of complex illusions that function as fruitful generators of scientific facts and theories. "Illusion" is not to be despised when it grounds such powerful truths.

Simian orientalism means that Western primatology has been about the construction of the self from the raw material of the other, the appropriation of nature in the production of culture, the ripening of the human from the soil of the animal, the clarity of white from the obscurity of color, the issue of man from the body of woman, the elaboration of gender from the resource of sex, the emergence of mind by the activation of body. To effect these transformative operations, simian "orientalist" discourse must first construct the terms: animal, nature, body, primitive, female. Traditionally associated with lewd meanings, sexual lust, and the unrestrained body, monkeys and apes mirror humans in a complex play of distortions over centuries of Western commentary on these troubling doubles. Primatology is Western discourse, and it is sexualized discourse. It is about potential and its actualization. Nature/culture and sex/gender are not loosely related pairs of terms; their specific form of relation is hierarchical appropriation, connected as Aristotle taught by the logic of active/passive, form/matter, achieved form/resource, man/animal, final/material cause. Symbolically, nature and culture, as well as sex and gender, mutually (but not equally) construct each other; one pole of a dualism cannot exist without the other.

Said's critique of orientalism should alert us to another important point: neither sex nor nature is the truth underlying gender and culture, any more than the "East" is really the origin and distorting mirror of the "West." Nature and sex are as crafted as their dominant "others." But their functions and powers are different. The task [here] is to participate in showing how the whole dualism is built, what the stakes might be in the architectures, and how the building might be redesigned. It matters to know precisely how sex and nature become natural-technical objects of knowledge, as much as it matters to explain their doubles, gender and culture. It is not the case that no story could be told without these dualisms or that they are part of the structure of the mind or language. For one thing, alternative stories within primatology exist. But these binarisms have been especially *productive* and especially *problematic* for constructions of female and race-marked bodies; it is crucial to see how the binarisms may be deconstructed and maybe redeployed.

It seems nearly impossible for those who produce natural sciences and comment on them for a living really to believe that there is no *given* reality beneath the inscriptions of science, no untouchable sacred center to ground and authorize an innocent and progressive order of knowledge. Maybe in the humanities there is no recourse from representation, mediation, story-telling, and social saturation. But the sciences succeed that other faulty order of knowledge; the proof is in their power to convince and reorder the whole world, not just one local culture. The natural sciences are the "other" to the human sciences, with their tragic orientalisms. But these pleas do not survive scrutiny.

The pleas of natural scientists do not convince because they are set up as the "other." The claims are predictable and seem plausible to those who make them because they are built into the taxonomies of Western knowledge and because social and psychological needs are met by the persistent voices of the divided knowledge of natural and human sciences, by this division of labor and authority in the production of discourses. But these observations about predictable claims and social needs do not reduce natural sciences to a cynical "relativism" with no standards beyond arbitrary power. Nor does my argument claim there is no world for which people struggle to give an account, no referent in the system of signs and productions of meanings, no progress in building better accounts within traditions of practice. That would be to reduce a complex field to one pole of precisely the dualisms under analysis, the one designated as ideal to some impossible material, appearance to some forbidden real.

The point of my argument is rather that natural sciences, like human sciences, are inextricably *within* the processes that give them birth. And so, like the human sciences, the natural sciences are culturally and historically specific, modified, involved. They matter to real people. It makes sense to ask what stakes, methods, and kinds of authority are involved in natural scientific accounts, how they differ, for example, from religion or ethnography. It does not make sense to ask for a form of authority that escapes the web of the highly productive cultural fields that make the accounts possible in the first place. The detached eye of objective science is an ideological fiction, and a powerful one. But it is a fiction that hides—and is designed to hide—how the powerful discourses of the natural sciences really work. Again, the limits are *productive,* not reductive and invalidating.

One grating consequence of my argument is that the natural sciences are legitimately subject to criticism on the level of "values," not just "facts." They are subject to cultural and political evaluation "internally," not just "externally." But the evaluation is also implicated, bound, full of interests and stakes, part of the field of practices that make meanings for real people accounting for situated lives, including highly structured things called scientific observations. The evaluations and critiques cannot leap over the crafted standards for producing credible accounts in the natural sciences because neither the critiques nor the objects of their discourse have any place to stand "outside" to legitimate such an arrogant overview. To insist on value and story-ladenness at the heart of the production of scientific knowledge is not equivalent to standing nowhere talking about nothing but one's biases—quite the opposite. Only the pose of disinterested objectivity makes "concrete objectivity" impossible.

Part of the difficulty of approaching the embedded, interested, passionate constructions of science nonreductively derives from an inherited analytical tradition, deeply indebted to Aristotle and to the transformative history of "White Capitalist Patriarchy" (how may we name this scandalous Thing?) that turns everything into a resource for appropriation. As "resource" an object of knowledge is finally only matter for the seminal power, the act, of the knower. Here, the object both guarantees and refreshes the power of the knower, but any status as *agent* in the productions of knowledge must be denied the object. It—the world—must, in short, be objectified as thing, not agent; it must be matter for the self-formation of the only social being in the productions of knowledge, the human knower. Nature is only the raw material of culture, appropriated, preserved, enslaved, exalted, or otherwise made flexible for disposal by culture in the logic of capitalist colonialism. Similarly, sex is only the matter to the act of gender; the productionist logic seems inescapable in traditions of Western binarisms. This analytical and historical narrative logic accounts for my nervousness about the sex/gender distinction in the recent history of feminist theory as a way to approach reconstructions of what may count as female and as nature in primatology—and why those reconstructions matter beyond the boundaries of primate studies. It has seemed all but impossible to avoid the trap of an appropriationist logic of domination built into the nature/culture binarism and its generative lineage, including the sex/gender distinction.

READING IN THE BORDERLANDS

There are many subjects in the history of biology and anthropology that could sustain the themes discussed [here], so why explore primate sciences in particular? The principal reason is that monkeys and apes, and human beings as their taxonomic kin, exist on the boundaries of so many struggles to determine what will count as knowledge. Primates are not nicely boxed into a specialized and secured discipline or field. Even in the late twentieth century, many kinds of people can claim to know primates, to the chagrin and dismay of many other contestants for official expertise. The cost of destabilizing knowledge about primates remains within reach not only for practitioners of several fields in the life and human sciences, but for people on the fringes of any science—like science writers, philosophers, historians, and zoo goers. In addition, storytelling about animals is such a deeply popular practice that the discourse produced within scientific specialties is appropriated by other people for their own ends. The boundary between technical and popular discourse is very fragile and permeable. Even in the late twentieth century, the language of primatology is accessible in contentious political debate about human nature, history, and futures. This remains true despite a transformation of specialized discourses in primatology into the language of mathematics, systems theories, ergonomic analysis, game theory, life history strategies, and molecular biology.

Some of the interesting border disputes about primates, who and what they are (and who and what they are for), are between psychiatry and zoology, biology and anthropology, genetics and comparative psychology, ecology and medical research, agriculturalists and tourist industries in the "third world," field researchers and laboratory scientists, conservationists and multinational logging companies, poachers and game wardens, scientists and administrators in zoos, feminists and antifeminists, specialists and lay people, physical anthropologists and ecological-evolutionary biologists, established scientists and new Ph.D.s, women's studies students and professors in animal behavior courses, linguists and biologists, foundation officials and grant applicants, science writers and researchers, historians of science and real scientists, Marxists and liberals, liberals and neoconservatives.

I hope that the readers who begin in the position of one of the intended audiences find themselves invited to become members of all of the audiences. And I hope that readers will not be "audience" in the sense of receivers of a finished story. Conventions within the narrative field of SF seem to require readers radically to rewrite stories in the act of reading them. My placing this account of primatology within SF—the narratives of speculative fiction and scientific fact—is an invitation for the readers of *Primate Visions*—historians, culture critics, feminists, anthropologists, biologists, antiracists, and nature lovers—to remap the borderlands between nature and culture. I want the readers to find an "elsewhere" from which to envision a different and less hostile order of relationships among people, animals, technologies, and land. I want to set new terms for the traffic between what we have come to know historically as nature and culture.

NOTES

1. John Varley's science fiction short story, "The Persistence of Vision," in *The Persistence of Vision* (New York: Dell, 1978), is part of the inspiration for *Primate Visions*. In the story, Varley constructs a utopian community designed and built by the deaf-blind. He then explores these people's technologies and other mediations of communication and their relations to sighted children and visitors. The interrogation of the limits and violence of vision is part of the politics of learning to revision.

2. Michel Foucault, *The Birth of the Clinic: An Archeology of Medical Perception*, trans. A. M. Sheridan Smith (New York: Pantheon, 1973); William R. Albury, "Experiment and Explanation of Bichat and Magendie," *Studies in the History of Biology* 1 (1977): 47–131; Georges Canguilhem, *On the Normal and Pathological*, trans. Carolyn R. Fawcett (Dordrecht: Reidel, 1978); Karl Figlio, "The Metaphor of Organization: An Historiographical Perspective on the Biomedical Sciences," *History of Science* 14 (1976): 17–53.

3. Teresa de Lauretis, "Signs of Wa/onder," in *The Technological Imagination*," ed. Teresa de Lauretis, A. Huyssen, and K. Woodward (Madison: Coda Press, 1980), p. 161.

4. Bruno Latour and Stephen Woolgar, *Laboratory Life: The Social Construction of Scientific Facts* (London: Sage, 1979), p. 213.

5. See also Bruno Latour, "Give Me a Laboratory and I Will Raise the World," in *Science Observed: Perspectives on the Social Study of Science* (London: Sage, 1983); Latour, *Science in Action: How to Follow Scientists and Engineers through Society* (Cambridge: Harvard University Press, 1987); Latour, *The Pasteurization of France*, trans. Alan Sheridan and John Law (Cambridge: Harvard University Press, 1988); W. Bijker et al., eds., *The Social Construction of Technological Systems: New Directions in the Sociology and History of Technology* (Cambridge: MIT Press, 1987): M. Callon and Bruno Latour, "Unscrewing the Big Leviathan, or How Do Actors Microstructure Reality? in *Advances in Social Theory and Methodology: Toward an Integration on Micro and Macro Sociologies*, ed. Karin Knorr-Cetina and A. Cicourel (London: Routledge & Kegan Paul, 1981); Karin Knorr-Cetina, "The Ethnographic Study of Scientific Work: Toward a Constructivist Interpretation of Science," in *Science Observed*; S. Traweek, *Beamtimes and Lifetimes: The World of High Energy Physics* (Cambridge: Harvard University Press, 1988).

6. Nancy Hartsock, "The Feminist Standpoint: Developing the

Ground for a Specifically Feminist Historical Materialism," in *Discovering Reality: Feminist Perspectives on Epistemology, Metaphysics, Methodology, and Philosophy of Science,* ed. Sandra Harding and Martha Hintikka (Dordrecht: Reidel, 1983); Sandra Harding, *The Science Question in Feminism* (Ithaca: Cornell University Press, 1986); Hilary Rose, "Head, Hands, and Heart: A Feminist Epistemology for the Natural Sciences," *Signs* 9 (1983): 73–90.

7. R. M. Young, "Science *Is* Social Relations," *Radical Science Journal* 5 (1977): 65–129; Young, "Is Nature a Labour Process?" in *Science, Technology, and Labour Process: Marxist Studies,* ed. L. Levidow and R. M. Young, vol. 2 (London: Free Associaltion Books, 1985); E. J. Voxen, "Life as a Productive Force: Capitalizing the Science and Technology of Molecular Biology," in *Science, Technology, and Labour Process;* Voxen, *The Gene Business: Who Should Control Biotechnology?* (New York: Harper and Row, 1983); Karl Figlio, "The Historiography of Scientific Medicine: An Invitation to the Human Sciences," *Comparative Studies in Society and History* 19 (1977): 262–86.

8. E. Fee, " Critiques of Modern Science: The Relationship of Feminism to Other Radical Epistemologies," in *Feminist Approaches to Science,* ed. R. Bleir (New York: Pergamon, 1986); Stephen J. Gould, *The Mismeasure of Man* (New York: W. W. Norton, 1981); E. Hammonds, "Race, Sex, and AIDS, the Construction of the 'Other,'" *Radical America* 6 (1986): 28–38; R. Hubbard et al., eds., *Biological Women—The Convenient Myth: A Collection of Essays and a Comprehensive Bibliography* (Cambridge, Mass.: Schenkman, 1982); S. L. Gilman, "Black Bodies, White Bodies: Toward an Iconography of Female Sexuality in Late Nineteenth Century Art, Medicine, and Literature," *Critical Inquiry* 12, no. 1 (1985): 204–42; M. Lowe and R. Hubbard, eds., *Women's Nature: Rationalization of Inequality* (New York: Pergamon, 1983); E. F. Keller, *Reflections on Gender and Science* (New Haven: Yale University Press, 1985).

9. G. Evelyn Hutchinson, *The Ecological Theater and the Evolutionary Play* (New Haven: Yale University Press, 1965).

10. Edward Said, *Orientalism* (New York: Pantheon, 1978).

11. Ibid., p. xiii.

NINE

REVIEW OF *PRIMATE VISIONS* BY DONNA HARAWAY

MATT CARTMILL

It is thus not simply false to say that Mallarme is a Platonist or a Hegelian. But it is above all not true. And vice versa.

Jacques Derrida, *Dissemination* (Chicago: University of Chicago Press, 1981), p. 207

Reprinted from Matt Cartmill, review of *Primate Visions: Gender, Race, and Nature in the World of Modern Science*, by Donna Haraway, *International Journal of Primathology* 12, no. 1 (1991): 67–75.

This is a book that contradicts itself a hundred times; but that is not a criticism of it, because its author thinks contradictions are a sign of intellectual ferment and vitality. This is a book that systematically distorts and selects historical evidence; but that is not a criticism, because its author thinks that all interpretations are biased, and she regards it as her duty to pick and choose her facts to favor her own brand of politics. This is a book full of vaporous, French-intellectual prose that makes Teilhard de Chardin sound like Ernest Hemingway by comparison; but that is not a criticism, because the author likes that sort of prose and has taken lessons in how to write it, and she thinks that plain, homely speech is part of a conspiracy to oppress the poor. This is a book that clatters around in a dark closet of irrelevancies for 450 pages before it bumps accidentally into its index and stops; but that is not a criticism, either, because its author finds it gratifying and refreshing to bang unrelated facts together as a rebuke to stuffy minds. This book infuriated me; but that is not a defect in it, because it is supposed to infuriate people like me, and the author would have been happier still if I had blown out an artery. In short, this book is flawless, because all its deficiencies are deliberate products of art. Given its assumptions, there is nothing here to criticize. The only course open to a reviewer who dislikes this book as much as I do is to question its author's fundamental assumptions—which are big-ticket items involving the nature and relationships of language, knowledge, and science.

Knowledge, says the proverb, is power, and this book exemplifies a school of thought that takes that proverb literally. In our culture, scientists are given power and prestige because they claim to know how the world works. Conversely, people who resent or fear scientists doubt that claim to knowledge and try to debunk it. There are always ample grounds for doubt. Anybody who practices science knows that it is hard to be objective; our fears and vanity and prejudices creep into our theories as readily as they enter into the speeches of congressmen or the predictions of astrologers. From that observation, it is only a step to the belief that scientists are nothing but politicians and shamans, and that objective knowledge is itself a myth cooked up by scientists to protect and enhance their power. That belief is the cornerstone of Donna Haraway's book.

The style of thinking and talking that Haraway has adopted in *Primate Visions* is what is sometimes called deconstructionist. She also refers to it as postmodernism or poststructuralism. This style is increasingly prevalent in the humanities and social sciences (where it dominates the Modern Language Association and many fields of social and cultural anthropology), and it is now beginning to be heard in archaeological circles.[1] In some quarters, it is regarded as an essential part of feminist consciousness. The deconstructionist mode can be summed up as a sophisticated skepticism rooted in a deep suspicion of ordinary language. It views plain speech as a Trojan horse full of secret biases that we cannot recognize or criticize if we insist on talking and thinking in a "plain," comfortable way. Deconstructionism derives chiefly from French models like Derrida and Foucault, whose much-imitated prose style—ironical, teasing, ambiguous, sibylline—is studiedly unlike ordinary language. The central tenet of deconstructionist thought, if anything so deliberately oblique can be said to have anything so straightforward as a tenet, is that all texts are subject to an infinite number of interpretations. From this it follows that it is neither interesting nor profitable to ask whether a particular text is "true." All claims to know the "truth" are, at bottom, really something else. What they are usually taken to be is assertions of power over others. As Foucault put it,

> It is not the activity of the subject of knowledge that produces a corpus of knowledge . . . but power-knowledge, the processes and struggles that traverse it and of which it is made up, that determines the forms and possible domains of knowledge.[2]

I take this to mean that politics, not empirical inquiry, determines what scientists are allowed to believe, though that may be putting it too baldly.

Donna Haraway quotes these words of Foucault's as an epigraph in this book, in which she attempts a deconstruction of primatology. A deconstruction of a text or concept is a reading that calls into question its underlying assumptions, its supposed objectivity, and its authority.

Deconstruction is not a friendly act, and Haraway's approach to science in general and to primatology in particular is an unfriendly one, which makes no effort to understand or to sympathize with the intentions of scientists. From the very first sentence of *Primate Visions*, Haraway makes it clear that what interests her about primate biology is not its ostensible subject matter. She has no interest in the objective "facts" about primates themselves, for the simple reason that there are none. Facts, reality, and nature are, in her eyes, constructs cooked up by Western scientific elites to justify and enhance their power over the oppressed—chiefly women, colonized third-world peoples, and the working class.

The questions Haraway asks in this book's first paragraph may convey some idea of her critical program, epistemology, and rhetorical style:

> How are love, power, and science intertwined in the construction of nature in the late twentieth century? . . . In what specific places, out of which social and intellectual histories, and with what tools is nature constructed as an object of erotic and intellectual desire? How do the terrible marks of gender and race enable and constrain love and knowledge in particular cultural traditions, including the modern natural sciences? Who may contest for what the body of nature will be?

Since facts and data are constructs, Haraway regards primatology and other sciences as literary forms. She acknowledges no epistemological difference between science and science fiction, between *Molecular Biology of the Gene* and *E.T.:* both are stories, differing only in features of narrative style. "Scientific practice may be considered a type of story-telling practice," Haraway writes. "The facts themselves are types of stories" (3–4). All those old positivist representations of science as cumulative knowledge grounded in observation are stories of another sort—"stories with a particular aesthetic, realism, and a particular politics, commitment to progress" (4). All statements about organisms necessarily take the form of stories:

> The discourse of biology, beginning near the first decades of the nineteenth century, has been about organisms, beings with a life history; i.e., a plot with structure and function. Biology is inherently historical, and its form of discourse is inherently narrative. Biology as a way of knowing the world is kin to Romantic literature, with its discourse about organic form and function. Biology is the fiction appropriate to objects called organisms; biology fashions the facts "discovered" from organic beings (4–5).

Accordingly, "*Primate Visions* reads the primate text as science fiction, where possible worlds are constantly reinvented in the contest for very real, present worlds" (5)—that is, for political power:

> Biology, and primatology, are inherently political discourses, whose chief objects of knowledge, such as organisms and ecosystems, are icons (condensations) of the whole of the history and politics of the culture that constructed them for contemplation and manipulation. The primate body itself is an intriguing kind of political discourse (10).

As one might expect from all this, the stories Haraway tells about primatology are also an intriguing kind of political discourse. Haraway disavows any intention of writing "a disinterested, objective study" herself, because "such studies are impossible for anyone" (3), and she makes her own political biases perfectly clear from the beginning. Her story depicts the history of primate biology as a contest over the body of nature, played out between the forces of good and evil. The evil cause is that of straight white males, capitalists, liberals, and individualists, who stand for "exploitation of the emergent Third World, obligatory and normative heterosexuality, masculine dominance of a progressively war-based scientific enterprise in industrial civilization . . . White Capitalist Patriarchy. . . . How may we name this scandalous Thing?" (13, 176). The good cause is that of those oppressed by the scandalous Thing. Western primatology originated in the 1930s as a scientific-mythical readout of "the structure of colonial discourse—that complex search for the primitive, authentic, and lost self, sought in the baroque dialectic between the wildly free and subordinated other" (245). White Capitalist Patriarchy dictated the terms of primatological discourse until the mid-1970s, when sociobiology came along and freed primatology from its male-centered paradigms. Nowadays, "primatology is a genre of feminist theory" (277).

The deconstructionist style is not very well suited to affirmation, and so Haraway has more to say about her patriarchal villains than about her feminist heroines. She begins with an attack on the American Museum of Natural History in its palmy days as a "Teddy Bear Patriarchy" (a double reference to stuffed animals and Teddy Roosevelt, consecrated to promulgating the myth of a pristine natural order endangered by the corrupting influences of culture and civilization). For reasons that never became entirely clear to me, she regards that myth as an instrument of the big oil companies (152, 185). Apparently the man-nature boundary is a construct that provides international corporations with a license to rape nature; when viewed as "wildly free and subordinated other," nature becomes a commodity. This central axiom of Haraway's viewpoint is supported only by bald assertions and ironic metaphors, and I was unconvinced by it. The major modern architects of the man-nature boundary—Rousseau, Wordsworth, Thoreau, Jeffers, and so on—seem like unlikely cheerleaders for industrial capitalism.

In succeeding chapters, Haraway deconstructs Robert Yerkes, C. R. Carpenter, and Stuart Altmann (who are depicted as concerned with developing techniques and rationales for maintaining White Male Capitalist control over workers and women), Jane Goodall (another instrument of Big Oil), S. L. Washburn (whose "new physical anthropology" is construed here as part of a neocolonial justification for White Male Capitalist domination of the third world), Harry Harlow (a phallocratic sadist acting out his misogynistic fantasies with monkeys), and other leading American students of primate behavior. There is some substance to most of Haraway's caricatures, and even those that (like the one of Washburn) strike me as completely wrong-headed are laced with provocative insights.

In part III of her book, headed "The Politics of Being Female," Haraway examines the work of women like Adrienne Zihlman and Linda Fedigan, whose politics she admires and sees reflected in their scientific writings. But since the only style of analysis she commands is deconstruction, she has no means of praising these scientists in their own terms. The best she can do is to try to make them out as sardonic deconstructionists like herself, interested less in understanding primates than in mocking and subverting the rationalist, gender-inscribed presuppositions of White Male Capitalism. "Laughter is an indispensable tool in deconstructions of the bio-politics of being female," she insists. "Suspicion and irony are basic to feminist reinscriptions of nature's text." When she can reasonably construe the writings of female primatologists as being ironic and subversive, Haraway hails them as fellow architects of a new consciousness. When she can no longer evade the suspicion that some of them are trying to discover truths about the order of nature, she is forced to put them down gently as dupes who have swallowed the patriarchal assumptions imbedded in the concepts of "truth," "order," and "nature." Haraway's deconstructionist language gets denser and more oracular when she criticizes female scientists—for example, when she chides Goodall for failing to promote a postmodernist sensibility at Gombe:

> What is too dim [in Goodall's work] is a dimension problematizing (not erasing) the mythic, scientific, and individual axes; i.e., the historical. By history I mean a corrosive sense of the contradiction and multiple material-semiotic processes at the heart of scientific knowledge. History . . . is a discipline reworked by postmodern insights about always split, fragmented, and multiple subjects, identities, and collectivities. All units and actors cohere partially and provisionally, held together by complex material-semiotic-social practices. In the space opened up by such contradictions and multiplicities lies the possibility for reflexive responsibility for the shape of narrative fields (172).

I don't think that whatever "dimness" Haraway finds in Goodall's work is much illuminated by these words.

There are real insights and intermittent flashes of brilliance scattered through this book, and all primatologists will benefit from reading it and getting their preconceptions shaken up. Haraway's challenging analyses of the social, political, and empirical factors that have induced and guided the growth of feminist ideas in contemporary primatology are worth the modest price of the book all by themselves. But the book's virtues are outweighed by the faults that arise from Haraway's postmodernist epistemology. The worst of these faults is her refusal ever to deal with the past on its own terms, to give an account of people's actions in terms of their own ideas and intentions. Because she is not really interested in the thought of the past, but only in poking holes in it to reveal the scandalous Thing lurking within, she does not hesitate to caricature it into unintelligibility, leaving out vast sectors of the primatological tradition and distorting others to make them fit her picture. This approach may be appropriate for Haraway, who believes that reality is an artifact constructed for political ends, but it makes it hard to take her seriously as a historian of ideas.

Haraway's "poststructuralist" approach to history is thoroughly structuralist in its endless suspicious search for unperceived connections concealed behind surface appearances. It accords with Lévi-Strauss's dictum that "understanding consists in the reduction of one type of reality to another; that true reality is never the most obvious of realities, and that . . . to reach reality we must first repudiate experience."[3] In *Primate Visions,* the search for occult understanding takes the form of an allusive play of suggestive juxtapositions hinting at underlying cultural themes too vast, complex, and portentous to be expressed in any less oblique way. Unfortunately, Haraway's juxtapositions often seem whimsical and gratuitous, like the supposed connection between stuffed animals and eugenics that she traces in the American Museum's African Hall:

> A hope is implicit in every architectural detail: in immediate vision of the origin, perhaps the future can be fixed. By saving the beginnings, the end can be achieved and the present can be tran-

> scended. . . . Restoration of the origin, the task of genetic hygiene, is achieved in Carl Akeley's African Hall by an art that began for him in the 1880s with the crude stuffing of P. T. Barnum's elephant, Jumbo, who had been run down by a railroad train, the emblem of the Industrial Revolution. The end of his task came in the 1920s, with his exquisite mounting of . . . the lone silverback male gorilla that dominates the diorama depicting the site of Akeley's own grave in the mountainous rain forest of the Congo, today's Zaire. So it could inhabit Akeley's monument to the purity of nature, this gorilla was killed in 1921, the same year the Museum hosted the Second International Congress of Eugenics. . . . Decadence—the threat of the city, civilization, machine—was stayed in the politics of eugenics and the art of taxidermy (26–27).

Is any of this really "implicit in every architectural detail"? I doubt it. I also doubt that the train that killed Jumbo is relevant to anything. These are flights of empty poetic fancy, and the whole connection that Haraway wants to draw between eugenics and taxidermy is really just as fanciful. It seems plausible only because the two things went on at the same time in the same building, and because eugenics can in some lights be seen as a backward-looking search for lost origins. But if the Eugenics Congress had been held at a Museum of Science and Technology, Haraway could have found equal significance in *that* conjunction, by describing eugenics as a forward-looking fantasy of racial progress through the new science of genetics, or some such. In fact, the connection between eugenics and taxidermy at the American Museum in the 1920s lay principally in the person of the museum's director, Henry Fairfield Osborn,[4] a haughty, egomaniacal, reactionary bigot about whom Haraway has practically nothing to say.

A poetic intelligence like Haraway's can always draw some sort of connection, however remote, between any two events whatever. But we are not obliged to take such connections seriously unless we are given some reason for thinking that they are not coincidental. Were other natural history museums of the period busy putting up stuffed-animal dioramas? Were those museums also centers of eugenics agitation? If so, then Haraway's perception of Akeley's art deserves some credence; if not, we can discount it as based on a single suggestive coincidence. Haraway has not bothered to test her perceptions in this way. Readers of this journal will recognize in this test the essence of the scientific method: trying to figure out how meaningful a conjunction is by seeing whether it recurs regularly in similar circumstances. I think this is an important difference, maybe the most important single difference, between science and story-telling. Stories say, "Once upon a time"; science says, "Whenever *x*, then *y*." Narration is declarative; science is subjunctive.

Despite Haraway's protestations to the contrary, *Primate Visions* strikes me as an expression of hostility and contempt, to the scientific enterprise in general and to primatologists in particular. Science is grounded in the belief that there is a real world and that, by studying it and experimenting with it, we can understand, predict, and control its phenomena. To dismiss this belief as "the aesthetic of realism," a literary convention adopted for political ends, amounts to saying that scientists do not really understand what they are doing, and if they did, they would stop doing science and start doing the sort of thing Haraway demands of Goodall. I think it is fair to describe this contention as hostile and contemptuous.

The contempt for science expressed in *Primate Visions* is not wholly undeserved. Many scientists have deluded themselves into believing that their concepts were given by observation and that their prejudices spoke with the voice of Nature. Theories born of such delusions have engendered a lot of wasted effort and pointless suffering. But we can judge the effort as wasted and the suffering as pointless only because we have reasons for thinking that the underlying theories are (at least relatively) defective; and we cannot find such reasons unless we have valid criteria for evaluating competing accounts of the world.

Haraway, too, apparently thinks that there ought to be such criteria, because she insists that her perspective "does not reduce natural science to a cynical relativism with no standards beyond arbitrary power" (12). Maybe so. But if there are any other standards that we can legitimately use for choosing between conflicting visions of the world, she never tells us what they are, and never uses any herself. Her own evaluations of scientific theories are rooted in just such a standard of arbitrary power: theories that uphold the powerful are deemed bad, whereas those that question reigning orthodoxies are good. It is not clear that this is really a practical standard to apply to theories about, say, renal physiology or polymer chemistry. Haraway herself questions "whether scientific analysis could ever be postmodernist" and wonders "What would stable, replicable, cumulative knowledge about non-units look like?" (309). To this question, she offers no answer.

"Facts," argues Haraway, "are always theory-laden; theories are value-laden; therefore facts are value-laden" (288). Even if we accept the premises of this syllogism, its conclusion does not follow, because it hinges on a pun. It is rather like saying, "This bus is full of skeptics; skeptics are full of doubt; therefore this bus is full of doubt." Buses do not contain skeptics in the same way that skeptics contain doubt, and facts do not contain theories in the same way that theories contain values. These things are not related to each other like concentric boxes. Facts and theories and values are often all tangled up with each other, but they do not usually entail each other in any simple logical way, and the entanglements between them are not usually obvious ones. They become discernible only at a higher and vaguer level of analysis, at what might be called the level of the surrounding cultural matrix. It is in general not possible to infer underlying values from an isolated factual claim. If facts were always value-laden, we could tell at least *something* about a person's values from any declarative utterance. Since we cannot always make such inferences, the value-ladenness of many factual assertions must be contingent, not logically necessary. Facts must therefore be, at least in principle, independent from values.

In reading Haraway's book, I often thought about another left-wing literary figure, George Orwell, who came to quite different conclusions about the relationship among language, oppression, and the construction of facts. Toward the end of Orwell's *1984*, the inquisitor, O'Brien, forces the hero to abandon his old-fashioned belief in an external reality that limits human power. "Reality," says O'Brien, "is inside the skull. . . . You must get rid of those nineteenth-century ideas about the laws of nature. We make the laws of nature." For certain purposes, says O'Brien, it is convenient to assume that the earth circles the sun; for other purposes, the reverse assumption is convenient. We can learn to accept either assumption or both at once if the Party demands it. "The stars can be near or distant, according as we need them. Do you suppose our mathematicians are unequal to that? Have you forgotten doublethink?" With O'Brien's help, the hero finally shakes off his belief in nature and reality and comes to understand how two twos can make five if the Party says so. I have the uneasy feeling that Haraway might, at least in principle, regard that liberation from the constraints of "fact" as an intellectual triumph.

In its denial of external reality as something given, its obsession with motifs of dominance and power, and its rejection of logical dualisms (war is peace, freedom is slavery, and what is untrue is above all not simply false, and vice versa), the postmodernist sensibility displayed in this book is strangely reminiscent of the official philosophy of Orwell's posttotalitarian state. Haraway is, of course, no propagandist for Big Brother, but she has chosen not to acknowledge a truth that Orwell, like Marx, always insisted on: that there is a world antecedent to human ambition and desire, and that the powerful and arrogant are occasionally constrained to acknowledge that objective reality by having their noses rubbed in it. To deny that reality is to deny that there are external constraints on human power. It amounts to saying that the right TV programs can keep the masses hypnotized forever, because there is nothing beyond the screen that might wake them up unexpectedly. Saying that may feel like a brave gesture of defiance to Haraway, but from where I sit it looks like a capitulation.

NOTES

1. Paul G. Bahn, "Motes and Beams: A Further Response to White on the Upper Paleolithic," *Current Anthropology* 31 (1990): 71–76.

2. Michel Foucault, *Discipline and Punish* (New York: Pantheon, 1979), p. 28.

3. Claude Lévi-Strauss, *Tristes Tropiques* (New York: Antheneum, 1964), pp. 61–62.

4. M. P. Sutphen, *W. K. Gregory's and H. F. Osborn's Conflict in the Galton Society and Physical Anthropology 1918–1935*, master's thesis, Duke University, Durham, North Carolina, 1988.

TEN

SOKAL'S HOAX AND SELECTED RESPONSES

STEVEN WEINBERG

NEW YORK REVIEW OF BOOKS, 8 AUGUST 1996

Like many other scientists, I was amused by news of the prank played by the NYU mathematical physicist Alan Sokal. Late in 1994 he submitted a sham article to the cultural studies journal *Social Text,* in which he reviewed some current topics in physics and mathematics, and with tongue in cheek drew various cultural, philosophical, and political morals that he felt would appeal to fashionable academic commentators on science who question the claims of science to objectivity.

The editors of *Social Text* did not detect that Sokal's article was a hoax, and they published it in the journal's spring/summer 1996 issue.[1] The hoax was revealed by Sokal in an article for another journal, *Lingua Franca*;[2] in which he explained that his *Social Text* article had been "liberally salted with nonsense," and in his opinion was accepted only because "(a) it sounded good and (b) it flattered the editors' ideological preconceptions." Newspapers and newsmagazines throughout the United States and Britain carried the story. Sokal's hoax may join the small company of legendary academic hoaxes, along with the pseudo-fossils of Piltdown man planted by Charles Dawson and the pseudo-Celtic epic *Ossian* written by James Macpherson. The difference is that Sokal's hoax served a public purpose, to attract attention to what Sokal saw as a decline of standards of rigor in the academic community, and for that reason it was unmasked immediately by the author himself.

The targets of Sokal's satire occupy a broad intellectual range. There are those "postmoderns" in the humanities who like to surf through avant-garde fields like quantum mechanics or chaos theory to dress up their own arguments about the fragmentary and random nature of experience. There are those sociologists, historians, and philosophers who see the laws of nature as social constructions. There are cultural critics who find the taint of sexism, racism, colonialism, militarism, or capitalism not only in the practice of scientific research but even in its conclusions. Sokal did not satirize creationists or other religious enthusiasts who in many parts of the world are the most dangerous adversaries of science,[3] but his targets were spread widely enough, and he was attacked or praised from all sides.

Entertaining as this episode was, from press reports I could not immediately judge what it proved. Suppose that, with tongue in cheek, an economist working for a labor union submitted an article to the *National Review,* giving what the author thought were false economic arguments against an increase in the statutory minimum wage. What would it prove if the article were accepted for publication? The economic arguments might still be cogent, even though the author did not believe in them.

I thought at first that Sokal's article in *Social Text* was intended to be an imitation of academic babble, which any editor should have recognized as such. But in reading the article I found that this is not the case. The article expresses views that I find surreal, but with a few exceptions Sokal at least makes it pretty clear what these views are. The article's title, "Transgressing the Boundaries: Toward a Transformative Hermeneutics of Quantum Gravity," is more obscure than almost anything in his text. (A physicist friend of mine once said that in facing death, he drew some consolation from the reflection that he would never again have to look up the word "hermeneutics" in the dictionary.) I got the impression that Sokal finds it difficult to write unclearly.

Where the article does degenerate into babble, it is not in what Sokal himself has written, but in the writings of the genuine postmodern cultural critics quoted by Sokal. Here, for instance, is a quote that Sokal takes from the oracle of deconstructionism, Jacques Derrida:

> The Einsteinian constant is not a constant, is not a center. It is the very concept of variability—it is, finally, the concept of the game. In other words, it is not the concept of *something*—of a center starting from which an observer could master the field—but the very concept of the game. [pt. 1, sec. 2]

I have no idea what this is intended to mean.

I suppose that it might be argued that articles in physics journals are also incomprehensible to the uninitiated. But physicists are forced to use a technical language, the language of mathematics. Within this limitation, we try to be clear, and when we fail we do not expect our readers to confuse obscurity with profundity. It never was true that only a dozen people could understand Einstein's papers on general relativity, but if it had been true, it would have been a failure of Einstein's, not a mark of his brilliance. The papers of Edward Witten of the Institute for Advanced Study at Princeton, which are today consistently among the most significant in the promising field of string theory, are notably easier for a physicist to read than most other work in string theory. In contrast, Derrida and other postmoderns do not seem to be saying anything that requires a special technical language, and they do not seem to be trying very hard to be clear. But those who admire such writings presumably would not have been embarrassed by Sokal's quotations from them.

Part of Sokal's hoax was in his description of developments in physics. Much of his account was quite accurate, but it was heavily adulterated with howlers, most of which would have been detected by any undergraduate physics major. One of his running jokes had to do with the word "linear." This word has a precise mathematical meaning, arising from the fact that certain mathematical relationships are represented graphically by a straight line.[4] But for some postmodern intellectuals, "linear" has come to mean unimaginative and old-fashioned, while "nonlinear" is understood to be somehow perceptive and avant-garde. In arguing for the cultural importance of the quantum theory of gravitation, Sokal refers to the gravitational field in this theory as "a noncommuting (and hence nonlinear) operator." Here "hence" is ridiculous; "noncommuting"[5] does not imply "nonlinear," and in fact quantum mechanics deals with things that are both noncommuting and linear.

Sokal also writes that "Einstein's equations [in the general theory of relativity] are highly nonlinear, which is why traditionally trained mathematicians find them so difficult to solve." The joke is in the words "traditionally trained"; Einstein's equations are nonlinear, and this does make them hard to solve, but they are hard for anyone to solve, especially someone who is not traditionally trained. Continuing with general relativity, Sokal correctly remarks that its description of curved space-time allows arbitrary changes in the space-time coordinates that we use to describe nature. But then he solemnly pronounces that "the π of Euclid and the G of

Newton, formerly thought to be constant and universal, are now perceived in their ineluctable historicity" [pt. 1, sec. 2]. This is absurd—the meaning of a mathematically defined quantity like pi cannot be affected by discoveries in physics, and in any case both pi and *G* continue to appear as universal constants in the equations of general relativity.

In a different vein, Sokal pretends to give serious consideration to a crackpot fantasy known as the "morphogenetic field." He refers to complex number theory as a "new and still quite speculative branch of mathematical physics," while in fact it is nineteenth-century mathematics and has been as well established as anything ever is. He even complains (echoing the sociologist Stanley Aronowitz) that all of the graduate students in solid-state physics will be able to get jobs in that field, which will be news to many of them.

Sokal's revelation of his intentional howlers drew the angry response that he had abused the trust of the editors of *Social Text* in his credentials as a physicist, a complaint made by both sociologist Steve Fuller and English professor Stanley Fish.[6] (Fish is the executive director of Duke University Press, which publishes *Social Text*, and is reputed to be the model for Morris Zapp, the master of the academia game in David Lodge's comic novels.) The editors of *Social Text* also offered the excuse that it is not a refereed journal, but a journal of opinion.[7] Maybe under these circumstances Sokal was naughty in letting the editors rely on his sincerity, but the article would not have been very different if Sokal's account of physics and mathematics had been entirely accurate. What is more revealing is the variety of physics and mathematics bloopers in remarks of others that Sokal slyly quotes with mock approval. Here is the philosopher Bruno Latour on special relativity:

> How can one decide whether an observation made in a train about the behavior of a falling stone can be made to coincide with the observation of the same falling stone from the embankment? If there are only one, or even *two*, frames of reference, no solution can be found.... Einstein's solution is to consider *three* actors...

This is wrong; in relativity theory there is no difficulty in comparing the results of two, three, or any number of observers. In other quotations cited by Sokal, Stanley Aronowitz misuses the term "unified field theory"; feminist theorist Luce Irigaray deplores mathematicians' neglect of spaces with boundaries, though there is a huge literature on the subject. The English professor Robert Markley calls quantum theory nonlinear, though it is the only known example of a precisely linear theory. And both philosopher Michael Serres (a member of the Académie Française) and arch-postmodernist Jean-François Lyotard grossly misrepresent the view of time in modern physics. Such errors suggest a problem not just in the editing practices of *Social Text,* but in the standards of a larger intellectual community.

It seems to me though that Sokal's hoax is most effective in the way that it draws cultural or philosophical or political conclusions from developments in physics and mathematics. Again and again Sokal jumps from correct science to absurd implications, without the benefit of any intermediate reasoning. With a straight face, he leaps from Bohr's observation that in quantum mechanics "a complete elucidation of one and the same object may require diverse points of view which defy a unique description" to the conclusion that "postmodern science" refutes "the authoritarianism and elitism inherent in traditional science." He blithely points to catastrophe theory and chaos theory as the sort of mathematics that can lead to social and economic liberation. Sokal shows that people really do talk in this way by quoting work of others in the same vein, including applications of mathematical topology to psychiatry by Jacques Lacan and to film criticism by Jacques-Alain Miller.

I find it disturbing that the editors of *Social Text* thought it plausible that a sane working physicist would take the positions satirized in Sokal's article. In their defense of the decision to publish it, the editors explain that they had judged that it was "the earnest attempt of a professional scientist to seek some sort of affirmation from postmodern philosophy for developments in his field."[8] In an introduction to the issue of *Social Text* in which Sokal's article appears, one of the editors mentions that "many famous scientists, especially physicists, have been mystics."[9] There may be some working physicists who are mystics, though I have never met any, and I can't imagine any serious physicist who holds views as bizarre as those that Sokal satirized. The gulf of misunderstanding between scientists and other intellectuals seems to be at least as wide as when C. P. Snow worried about it three decades ago.

After Sokal exposed his hoax, one of the editors of *Social Text* even speculated that "Sokal's parody was nothing of the sort, and that his admission represented a change of heart, or a folding of his intellectual resolve."[10] I am reminded of the case of the American spiritualist Margaret Fox. When she confessed in 1888 that her career of séances and spirit rappings had all been a hoax, other spiritualists claimed that it was her confession that was dishonest.

Those who seek extrascientific messages in what they think they understand about modern physics are digging dry wells. In my view, with two large exceptions, the results of research in physics (as opposed, say, to psychology) have no legitimate implications whatever for culture or politics or philosophy. (I am not talking here about the technological applications of physics, which of course did have a huge effect on our culture, or about its use as metaphor, but about the direct logical implications of purely scientific discoveries themselves.) The discoveries of physics may become relevant to philosophy and culture when we learn the origin of the universe or the final laws of nature, but not for the present.

The first of my exceptions to this statement is jurisdictional: discoveries in science sometimes reveal that topics like matter, space, and time, which had been thought to be proper subjects for philosophical argument, actually belong in the province of ordinary science. The other, more important, exception to my statement is the profound cultural effect of the discovery, going back to the work of Newton, that nature is strictly governed by impersonal mathematical laws. Of course, it still remains for us to get the laws right, and to understand their range of validity; but as far as culture or philosophy is concerned the difference between Newton's and Einstein's theories of gravitation or between classical and quantum mechanics is immaterial.

There is a good deal of confusion about this, because quantum mechanics can seem rather eerie if described in ordinary language. Electrons in atoms do not have definite positions or velocities until these properties are measured, and the measurement of an electron's velocity wipes out all knowledge of its position. This eeriness has led Andrew Ross, one of the editors of *Social Text*, to remark elsewhere that "quantitative rationality—the normative description of scientific materialism—can no longer account for the behavior of matter at the level of quantum reality."[11] This is simply wrong. By rational processes today we obtain a complete quantitative description of atoms in terms of what is called the wave function of the atom.[12] Once one has calculated the wave function, it can be used to answer any question about the energy of the atom or its interaction with light. We have replaced the precise Newtonian language of particle trajectories with the precise quantum language of wave functions, but as far as quantitative rationality is concerned, there is no difference between quantum mechanics and Newtonian mechanics.

I have to admit at this point that physicists share responsibility for the widespread confusion about such matters. Sokal quotes some dreadful examples of Werner Heisenberg's philosophical wanderings, as for instance: "Science no longer confronts nature as an objective observer, but sees itself as an actor in this interplay between man [*sic*] and nature." (Heisenberg was one of the great physicists of the twentieth century, but he could not always be counted on to think carefully, as shown by his technical mistakes in the German nuclear weapons program.)[13] More recently scientists like Ilya Prigogine[14] have claimed a deep philosophical significance for work on nonlinear dynamics,[15] a subject that is interesting enough without the hype.

So much for the cultural implications of discoveries in science. What of the implications for science of its cultural and social context? Here scientists like Sokal find themselves in opposition to many sociologists, historians, and philosophers as well as postmodern literary theorists. In this debate, the two sides often seem to be talking past each other. For instance, the sociologists and historians sometimes write as if scientists had not learned anything about the scientific method since the days of Francis Bacon, while of course we know very well how complicated the relation is between theory and experiment, and how much the work of science depends on an appropriate social and economic setting. On the other hand, scientists sometimes accuse others of taking a completely relativist view, of not believing in objective reality. With dead seriousness, Sokal's hoax cites "revisionist studies in the history and philosophy of science" as casting doubt on the post-Enlightenment dogma that "there exists an external world, whose properties are independent of any individual human being and indeed of humanity as a whole" [pt. 1, intro.]. The trouble with the satire of this particular message is that most of Sokal's targets deny that they have any doubt about the existence of an external world. Their belief in objective reality was reaffirmed in response to Sokal's hoax both in a letter to the *New York Times* by the editors of *Social Text*[16] and in the Op-Ed article by Stanley Fish [pt. 3, "Professor Sokal's Bad Joke"].

I don't mean to say that this part of Sokal's satire was unjustified. His targets often take positions that seem to me (and I gather to Sokal) to make no sense if there is an objective reality. To put it simply, if scientists are talking about something real, then what they say is either true or false. If it is true, then how can it depend on the social environment of the scientist? If it is false, how can it help to liberate us? The choice of scientific question and the method of approach may depend on all sorts of extrascientific influences, but the correct answer when we find it is what it is because that is the way the world is. Nevertheless, it does no good to satirize views that your opponent denies holding.

I have run into the same sort of stumbling block myself. In an early draft of my book *Dreams of a Final Theory,*[17] I criticized the feminist philosopher of science, Sandra Harding (a contributor to *Social Text*), for taking a relativist position that denied the objective character of physical laws. In evidence I quoted her as calling modern science (and especially physics) "not only sexist but also racist, classist, and culturally coercive," and arguing that "physics and chemistry, mathematics and logic, bear the fingerprints of their distinctive cultural creators no less than do anthropology and history."[18] It seemed to me that this statement could make sense only to a relativist. What is the good of claiming that the conclusions of scientific research should be friendlier to multicultural or feminist concerns if these conclusions are to be an accurate account of objective reality? I sent a draft of this section to Harding, who pointed out to me various places in her writing where she had explicitly denied taking a relativist position. I took the easy way out; I dropped the accusation of relativism, and left it to the reader to judge the implications of her remarks.

Perhaps it would clarify what is at issue if we were to talk not about whether nature is real but about the more controversial question, whether scientific knowledge in general and the laws of physics in particular are real.

When I was an undergraduate at Cornell I heard a lecture by a professor of philosophy (probably Max Black) who explained that whenever anyone asked him whether something was real, he always gave the same answer. The answer was "Yes." The tooth fairy is real, the laws of physics are real, the rules of baseball are real, and the rocks in the fields are real. But they are real in different ways. What I mean when I say that the laws of physics are real is that they are real in pretty much the same sense (whatever that is) as the rocks in the fields, and not in the same sense (as implied by Fish)[19] as the rules of baseball. We did not create the laws of physics or the rocks in the field, and we sometimes unhappily find that we have been wrong about them, as when we stub our toe on an unnoticed rock, or when we find we have made a mistake (as most physicists have) about some physical law. But the languages in which we describe rocks or in which we state physical laws are certainly created socially, so I am making an implicit assumption (which in everyday life we all make about rocks) that our statements about the laws of physics are in a one-to-one correspondence with aspects of objective reality. To put it another way, if we ever discover intelligent creatures on some distant planet and translate their scientific works, we will find that we and they have discovered the same laws.

There is another complication here, which is that none of the laws of physics known today (with the possible exception of the general principles of quantum mechanics) are exactly and universally valid. Nevertheless, many of them have settled down to a final form, valid in certain known circumstances. The equations of electricity and magnetism that are today known as Maxwell's equations are not the equations originally written down by Maxwell; they are equations that physicists settled on after decades of subsequent work by other physicists, notably the English scientist Oliver Heaviside. They are understood today to be an approximation that is valid in a limited context (that of weak, slowly varying electric and magnetic fields), but in this form and in this limited context they have survived for a century and may be expected to survive indefinitely. This is the sort of law of physics that I think corresponds to something as real as anything else we know. On this point, scientists like Sokal and myself are apparently in clear disagreement with some of those whom Sokal satirizes. The objective nature of scientific knowledge has been denied by Andrew Ross[20] and Bruno Latour[21] and (as I understand them) by the influential philosophers Richard Rorty and the late Thomas Kuhn,[22] but it is taken for granted by most natural scientists.

I have come to think that the laws of physics are real because my experience with the laws of physics does not seem to me to be very different in any fundamental way from my experience with rocks. For those who have not lived with the laws of physics, I can offer the obvious argument that the laws of physics as we know them work, and there is no other known way of looking at nature that works in anything like the same sense. Sarah Franklin (in an article in the same issue of *Social Text* as Sokal's hoax) challenges an argument of Richard Dawkins that in relying on the working of airplanes we show our acceptance of the working of the laws of nature, remarking that some airlines show prayer films during takeoff to invoke the aid of Allah to remain safely airborne.[23] Does Franklin think that Dawkins's argument does not apply to her? If so, would she be willing to give up the use of the laws of physics in designing aircraft, and rely on prayers instead?

There is also the related argument that although we have not yet had a chance to compare notes with creatures on a distant planet, we can see that on Earth the laws of physics are understood in the same way by scientists of every nation, race, and—yes—gender. Some of the commentators on science quoted by Sokal hope that the participation of women or victims of imperialism will change the character of science; but as far as I can see, women and Third-World physicists work in just the same way as Western white male physicists do. It might be argued that this is just a sign of the power of entrenched scientific authority or the pervasive influence of Western society, but these explanations seem unconvincing to me. Although natural science is intellectually hegemonic, in the sense that we have a clear idea of what it means for a theory to be true or false, its operations are not socially hegemonic—authority counts for very little.

From time to time distinguished physicists who are past their best years, like Heisenberg in Germany in the 1950s or De Broglie in France, have tried to force physics in the direction of their own ideas; but where such mandarins succeed at all, it is only in one country, and only for a limited time. The direction of physics today is overwhelmingly set by young physicists, who are not yet weighed down with honors or authority, and whose influence—the excitement they stir up—derives from the objective progress that they are able to make. If our expression of the laws of nature is socially constructed, it is constructed in a society of scientists that evolves chiefly through grappling with nature's laws.

Some historians do not deny the reality of the laws of nature, but nevertheless refuse to take present scientific knowledge into account in describing the scientific work of the past.[24] This is partly to avoid anachronisms, like supposing that scientists of the past ought to have seen things in the way we do, and partly out of a preoccupation with maintaining the intellectual independence of historians.[25] The problem is that in ignoring present scientific knowledge, these historians give up clues to the past that cannot be obtained in any other way. In the late 1890s, J. J. Thomson carried out a celebrated series of measurements of the ratio of the electron's mass and charge, and though the values he found were spread over a wide range, he persistently emphasized measurements that gave results at the high end of the range. The historical record alone would not allow us to decide whether this was because these results tended to confirm his first measurement, or because these were actually more careful measurements. Why not use the clue that the second alternative is unlikely because the large value that was favored by Thomson is almost twice what we know today as the correct value?

A historian of science who ignores our present scientific knowledge seems to me like a historian of U.S. military intelligence in the Civil War who tells the story of McClellan's hesitations in the Virginia peninsula in the face of what McClellan thought were overwhelming Confederate forces without taking into account our present knowledge that McClellan was wrong. Even the choice of topics that attract the interest of historians has to be affected by what we now know were the paths that led to success. What Herbert Butterfield called the Whig interpretation of history is legitimate in the history of science in a way that it is not in the history of politics or culture, because science is cumulative, and permits definite judgments of success or failure.

Sokal was not the first to address these issues,[26] but he has done a great service in raising them so dramatically. They are not entirely academic issues, in any sense of the word "academic." If we think that scientific laws are flexible enough to be effected by the social setting of their discovery, then some may be tempted to press scientists to discover laws that are more proletarian or feminine or American or religious or Aryan or whatever else it is they want. This is a dangerous path, and more is at stake in the controversy over it than just the health of science. As I mentioned earlier, our civilization has been powerfully affected by the discovery that nature is strictly governed by impersonal laws. As an example I like to quote the remark of Hugh Trevor-Roper, that one of the early effects of this discovery was to reduce the enthusiasm for burning witches. We will need to confirm and strengthen the vision of a rationally understandable world if we are to protect ourselves from the irrational tendencies that still beset humanity.

SOKAL'S HOAX: AN EXCHANGE

New York Review of Books, 3 October 1996

To the Editors:

Alan Sokal's hoax is rapidly ceasing to be funny. An enterprise that originally had all the marks of a good joke is beginning to bring out the worst in respondents. This was obvious first in Sokal's own account in *Lingua Franca* of what he had done (very amusing) and why (tedious and self-righteous), and then in the firestorm of letters in the *New York Times* and various professional journals. But as teachers of a course on literature and science at Yale, we found Steven Weinberg's response to the hoax particularly troubling.

We do not, of course, wish to defend the shoddy scholarship of the *Social Text* editors, and we deplore the panculturalist views of those whom Weinberg attacks. But Weinberg has gone to the other extreme. If, in what follows, we concentrate on his views, it is because as a distinguished scientist, his intervention has the capacity to create mischief far beyond that of *Social Text*. Culture is too important to be left to *soi-disant* cultural critics, but it is also the case that Nobel Prize–winning physicists should not go unchallenged when they pronounce on culture—or science.

Alan Sokal and other scientists like Weinberg who have declared him a hero share one important feature with the editors of *Social Text*: both sides wish to locate science in a particular relation to other aspects of culture. The *Social Text* tribe sees science as merely a subfunction of the covering category "culture," while Weinberg flatly states, "The discoveries of physics may become relevant to philosophy and culture when we learn the origin of the universe or the final laws of nature, but not for the present." The *Social Text* editors fail to grant science sufficient distinctiveness in their homogenizing zeal. And Weinberg errs in the other direction: he argues science has *no* connection to the rest of culture. Both sides are guilty of egregious overstatement and impatiently exclude a middle where the real complexities are to be found.

Social Text and Weinberg both get the relation of science and culture wrong, but they do so in different ways. The claims of the *Social Text* editors have been self-discredited, not only by their acceptance of the Sokal piece, but by subsequent attempts they have made in statements to the press and in a *Lingua Franca* article to explain away their simple lack of basic seriousness, not only as scholars, but as intellectuals. Of Weinberg's seriousness, however, there can be no doubt. For this reason his argument is ultimately the more dangerous of the two, not only because it has a certain superficial plausibility (especially when presented in such clear prose and with so many entertaining examples), but because it represents in highly reductive terms a view probably held by many other scientists.

Such believers hold that science is an undertaking fundamentally different from other human activities for a number of reasons, but primarily because of its relation to a reality that is ultimate in the sense that its truths cannot be reduced to any other form of explanation. Such truths are objective, impersonal laws (a phrase Weinberg repeats like a mantra). Insofar as it is universal and extrahistorical, science is qualitively distinct from the rest of culture: science *is* nature, and therefore the very opposite of culture.

The most striking feature of this argument is its radical dualism: on the one side are timeless laws and selfless truths; on the other, is the social world of culture, with its ineluctable contingency, its ramifying particularity, its dictates that change with time. But the poles of this opposition are not equally weighted. Despite Weinberg's claim that in natural science "authority counts very little," his remarks are clearly intended to be definitive: Weinberg clearly sees himself as giving voice to the impersonal laws of natural science. These two fundamental aspects of Weinberg's argument—its obsessive dualism and its assumption of overwhelming authority—are grounded in a mode of thought frequently encountered in traditional societies where the distinctive feature of the culture is the dualism that separates the world of the profane from the sacred.

It would be absurd to compare the erudite and cosmopolitan

scientist to a member of, say, one of the tribes of central Australia described by Durkheim in *The Elementary Forms of Religious Life* were it not for the binarism that so compulsively attends the thinking of both: "The real characteristic of religious phenomena is that they always suppose a bipartite division of the whole universe, known and knowable, into two classes which embrace all that exists, but which radically exclude each other."

What does Weinberg's dualism include—and exclude? The innermost sanctum of his temple (before which all else is, in the etymological sense of the word, *profane*) is occupied by particle physicists. His use of the covering term "science" is deceptive, for it excludes microbiology, genetics, and the new brain sciences, to name merely a few. "Science" boils down to the work being done by a relatively small number of men in theoretical and high-energy physics. Outside the temple would be found first of all the enemies of the truth Weinberg specifically attacks for their impurity—the *Social Text* editors, of course, but as well most other historians, sociologists, and philosophers of science: "Scientists like Sokal [among whom Weinberg clearly counts himself] find themselves in opposition to many sociologists, historians, and philosophers as well as postmodern literary theorists." But in another circle of darkness would be found even other scientists, heretics such as Heisenberg and de Broglie. This list of apostates can be extended, by the same logic, to include the Newton who in his old age said, "I do not know what I may appear to the world; but to myself I seem to have been only like a boy playing on the seashore, and diverting myself now and then finding a smoother pebble or a prettier shell than ordinary, whilst the great ocean of truth lay all undiscovered before me."

Why, we may reasonably ask, has the mantle of purity Weinberg assumes in his attack on the editors of *Social Text* fallen precisely on the shoulders of particle physicists? Is it because defending pure science against philistine nonscientists gives particle physicists the opportunity to define science in their own image? The particle physicist stands ready in his role as pre-Kantian shaman to declare taboo all that which falls outside his narrow definitions. He defends pure science from the philistine laity by defining "science" in his own image. But it is not necessary to accept the mantras of particle physicists, with their reductionist view of science, to reject the foolishness of *Social Text*.

Since at least the pre-Socratics the greatest minds have striven to understand the relation of the incommensurable to our situatedness at a particular moment in a specific time. Since at least the Enlightenment this problem has been articulated in questions about how we relate physical sensation to processes of the understanding. All the gains we have made in this inclusive endeavor that has given new power not only to the natural sciences, but to other aspects of culture such as poetry and art as well, are now endangered by new polarizations in the science wars that rage about us. If we are to preserve a more holistic view of nature and our own place in it, we must resist not only those extremists who exclude formal knowledge in the name of a homgenizing concept of culture, but those as well who make equally privative claims for an immaculate conception of science. Or as Lionel Trilling deduced from the Leavis-Snow controversy, what gets lost in such conflicts between extremists is that quality of mind which creates the culture they claim to protect.

Michael Holquist
Professor of Comparative Literature
Yale University
New Haven, Connecticut

Robert Shulman
Sterling Professor of Molecular Biophysics
and Biochemistry
Yale University
New Haven, Connecticut

To the Editors:

With all the broad discussion of the Sokal affair, not enough has been made of the way the counterattackers against "postmodernism" and science studies get irrational and unscientific in the name of science and with the voice of common sense. Since part of the project of the writers Sokal mocks and Steven Weinberg criticizes is to force awareness of the metaphysical assumptions embedded in the language of common sense, they will often, even when sensible, sound obscure and irrational. I hope nobody wants to defend the awful jargon of much current theory and criticism. But Weinberg's demonstration of Derrida's vacuity on the basis of his failure to make sense of a passage selected by Sokal for mockery has no more weight than would a similar comment on science from a distinguished literary critic. Like Derrida or not, his linguistic project can't be dismissed with a commonsensical "I don't understand it."

Moreover, whereas every writer knows that all written words escape authorial control, Weinberg claims that the conclusions of physics can have no cultural implications. This is an extraordinary, a profoundly irrational claim. "Those who seek extrascientific messages in what they think they understand about modern physics," says Weinberg, "are digging dry wells." This is special pleading with a vengeance and Weinberg even provides two counterexamples.

How special the pleading is can be suggested by making an example of material from Sokal's parody. Sokal, Weinberg chortles, "leaps from Bohr's observation that in quantum mechanics 'a complete elucidation of one and the same object may require diverse points of view which defy a unique description' to the conclusion that 'postmodern science' refutes 'the authoritarianism and elitism inherent in traditional science.' " Weinberg laughs at the obvious absurdity of the conclusion, and then concludes that ANY cultural inference from Bohr's argument is illegitimate. Sure, Sokal probably had some real targets among cultural critics; but that an absurd inference is drawn doesn't for a moment preclude the possibility that there are other more reasonable ones. It is difficult *not* to see Bohr's argument as loaded with telling cultural implications. If Weinberg is right that such ideas apply always and everywhere, Bohr's observation ought to change the way lay people look at the world. Denying such possibility sounds more irrational, ideological, and misguided than anything *Social Text* did when it gave Sokal too much trust. It was not after all *Social Text* that drew the absurd conclusion about authoritarianism and elitism; it was Sokal, trying to ventriloquize work he didn't fully understand.

The new counteraggression of scientists hostile to "postmodernism" is surely the consequence of an economic pinch hurting them as well as humanists and social scientists. On all sides intellectual activity—not for profit or salvation—is under pressure. Both sides have got far too defensive and should be taking the awkward moment as an occasion for recognizing common interests and arguing their positions rationally. The most dangerous thing about Weinberg's kind of response is that it closes doors. He waves the banner of common sense, a banner that has been held higher, and waved more effectively by ideologues and demagogues, and in the vanguard of a war that inhibits science and crushes cultural critique.

George Levine
Center for the Critical Analysis of Contemporary Culture
Rutgers University
New Brunswick, New Jersey

To the Editors:

Steven Weinberg has used Alan Sokal's entertaining hoax (perpetrated against the editors of the formerly obscure journal *Social Text*) as an occasion for airing his own view on the nature of science, which he represents as those of "scientists" in contrast to "others" (historians, philosophers, and sociologists). He hopes thereby to "strengthen the vision of a rationally understandable world" so that we can "protect ourselves from the irrational tendencies that still beset humanity," for he is convinced that the "others" are promoting irrationality with their cultural relativism and attendant denial of objective reality. The difficulty with this representation is that all historians, philosophers, and sociologists that I know share Weinberg's hope for rational understanding and that none of them deny objective reality as Weinberg presents it. As a historian of physics I will remark on the relation between reality and relativism.

To evoke objective reality, Weinberg says that the laws of physics are as real as rocks and that we did not create them. Recognizing, however, that we have no access to any laws that we have not ourselves expressed, he retreats to the mundane formulation that our lawlike statements about aspects of the world correspond consistently to our experience of those aspects. Such laws he epitomizes by Maxwell's equations of electromagnetism, stressing that they are "an approximation that is valid in a limited context" or better, "valid in certain known circumstances." This validity criterion for realism—if it works it's real—would convert every empirically adequate simulation into reality and every antirealist into a realist. If such validity constitutes objective reality, Weinberg will search in vain for any historian who has ever denied the objective reality of Maxwell's equations. (His citation among others of Thomas Kuhn—so recently deceased after a lifelong attempt to articulate the nature of objectivity and rationality in science, including rigorous graduate courses in the history of electromagnetism—is outrageous.)

The issue of cultural relativism is not validity; it concerns multiplicity: the multiplicity of valid positions that have been available at any time, the many ways in which those positions have been embedded in the cultures from which they emerged, and the diverse processes through which they have both crossed cultural boundaries and have changed fundamentally over time, with massive implications for both science and the rest of culture. These objective realities leave contentious historians in a somewhat relativistic position as compared with such absolutist claims as Weinberg's that "as far as the culture of philosophy is concerned the difference between Newton's and Einstein's theories of gravitation or between classical and quantum mechanics is immaterial." This strange notion denied the experience of virtually every physicist who lived through the changes from 1900 to 1930. Continuing, Weinberg repeats the standard Copenhagen interpretation of quantum mechanics—sometimes called observer-created reality—according to which "electrons in atoms do not have definite positions or velocities until these properties are measured." While Einstein, among others, regarded this interpretation as unacceptable subjectivism, Weinberg presents it as an objective reality of no great historical significance, without telling us that it involved a renunciation (Bohr's term) of the classical causal description in space and time of the trajectory of a particle and without telling us that it has once again become controversial with the revival of David Bohm's long-dismissed deterministic (and holistic) alternative. Is Weinberg's blatant falsification of history excusable in the interest of damning the literary critic Andrew Ross as "simply wrong" in his remarks about the cultural significance of quantum mechanics? Multiplying errors, unlike multiplying interpretations, does not lead to wisdom.

There is a deeper question. Is it justifiable to try to expunge all mystical physicists (whom Weinberg has "never met"), "creationists and other religious enthusiasts" from the historical processes that have contributed to the immense power of science to predict and control events in the material world? Consider Maxwell's equations. They emerged during the course of the nineteenth century from the work of some of the most deeply religious people who have ever contemplated a battery: Oersted, discoverer of electromagnetism and author of *The Soul in Nature;* Faraday, devout member of the Sandemanian sect who discovered electromagnetic induction and articulated field theory; William Thomson (Lord Kelvin) and Maxwell, who mathematized field theory while holding that mathematical physics could produce only idealized accounts, accessible to finite human minds, of limited aspects of God's infinite action in the world; and a whole bevy of more spiritualist physicists who saw the electromagnetic field as the carrier of their dreams of uniting religion and science. These are some of the objective realities that historians of physics must learn to understand, however much we may disdain their role in contemporary society, for they are crucial aspects of the work in physics of the people who made Maxwell's equations. Again, the issue is not the validity of field theory (nor of the long-competing theories of action at a distance using retarded potentials) but of culturally embedded meanings.

And it does no good to suppose that science is no longer like that. Wolfgang Pauli, certainly one of the best mathematical physicists of the twentieth century, interpreted the oft-cited mysticism of Johannes Kepler, concerning the harmonies of planetary motion, in terms of his own belief in Jungian archetypes, which promised to account for both mathematical forms and his own dreams during years of analysis in the 1930s under Jungian guidance. More radically, Pascual Jordan, another of the primary mathematical founders of quantum mechanics and quantum field theory, presented these theoretical accomplishments during the thirties as a foundation not only for telepathy and clairvoyance but for Nazi politics as well. The only surprise is that (so far as I know) no physicist has yet presented the beautiful mathematical work of Edward Witten and others on string theory as the best-ever form of Platonic mysticism, since no direct empirical test seems conceivable at present.

In short, Weinberg presents us with an ideology of science, an ideology which radically separates science from culture, scientists from "others," and splits the personalities of rational and irrational components. However desirable this ideology may be in other respects, it will never do for comprehending the history of science. To preserve it, he has to indulge in the unsavory rhetorical ploy of dismissing other Nobelists, who have regarded their physics as having considerable philosophical and cultural significance, as aged oddballs appearing "from time to time" and "past their best." Thus Heisenberg's discussions of the subject-object problem become "philosophical wanderings" from someone who "could not always be counted on to think carefully." But scientists who have drawn on their philosophical, political, economic, and other beliefs for conceptual resources and motivation in pursuing their best scientific work permeate the history of physics. To remove them from quantum mechanics would be to wipe out the field: Planck, Bohr, de Broglie, Heisenberg, Pauli, Jordan, Schrödinger,

Weizsacker, to list only the obvious. So what is Weinberg up to? Is he not promoting a cultural agenda of his own in his attempt to rewrite history? Historical realities can be interpreted with validity in various ways, but like physical laws and rocks, they resist when kicked.

M. Norton Wise
Program in History of Science
Princeton University
Princeton, New Jersey

To the Editors:

Sokal's hoax and Weinberg's article explaining and amplifying its message effectively remove the smoke and mirrors from those social critics, philosophers, and historians of science who want to regard the human circumstances of a scientific discovery as more important than the discovery itself. As working physicists we know that the laws of nature we study are apprehended—tested and validated by independent experiments the same by women and men, and by people in every culture. The scientific revolution and continued discoveries give knowledge about the universe that can be comprehended and used by all. We commend Sokal and Weinberg for their defense of the scientific revolution.

Nina Byers
Professor of Physics
University of California at Los Angeles
Los Angeles, California

Claudio Pellegrini
Professor of Physics
University of California at Los Angeles
Los Angeles, California

STEVEN WEINBERG REPLIES

I am grateful to those who sent comments on my article "Sokal's Hoax," including those who, by disagreeing with me, have given me this chance to take another whack at the issues it raised.

Professors Holquist and Shulman have me dead to rights in calling my views dualist. I think that an essential element needed in the birth of modern science was the creation of a gap between the world of physical science and the world of human culture.[27] Endless trouble has been produced throughout history by the effort to draw moral or cultural lessons from discoveries of science. The physics and biology of Aristotle were largely based on a conception of naturalness, which was believed also to have moral and cultural implications, as for instance that some people are naturally slaves. After relativity theory became widely publicized in 1919, the archbishop of Canterbury like many others conscientiously worried over the effect that relativity was going to have on theology, and had to be reassured by Einstein himself in 1921 that relativity had no implications for religion.[28] Professors Holquist and Shulman quote Durkheim for the proposition that a gap between ways of viewing reality such as that between science and culture is characteristic of religious phenomena, but I think that just the opposite is true; if you want to find astronomy all muddled with cultural or moral values, you would turn to Dante's *Paradiso* rather than Galileo's *Dialogo*. In trying to revive a "holistic view of nature," Professors Holquist and Shulman are seeking to fill a much-needed gap.

Quantum mechanics provides a good example of the need to maintain this separation between physics and other forms of culture. Quantum mechanics has been variously cited as giving support to mysticism, or free will, or the decline of quantitative rationality. Now, I would agree that anyone is entitled to draw any inspiration they can from quantum mechanics, or from anything else. This is what I meant when I wrote that I had nothing to say against the use of science as metaphor. But there is a difference between inspiration and implication, and in talking of the "telling cultural implications" of quantum mechanics, Professor Levine may be confusing the two. There is simply no way that any cultural consequences can be *implied* by quantum mechanics. It is true that quantum mechanics does "apply always and everywhere," but what applies is not a proverb about diverse points of view but a precise mathematical formalism, which among other things tells us that the difference between the predictions of quantum mechanics and prequantum classical mechanics, which is so important for the behavior of atoms, becomes negligible at the scale of human affairs.

I suggest the following thought experiment. Suppose that physicists were to announce the discovery that, beneath the apparently quantum mechanical appearance of atoms, there lies a more fundamental substructure of fields and particles that behave according to the rules of plain old classical mechanics. Would Professor Levine find it necessary to rethink his views about culture of philosophy? If so, why? If not, then in what sense can these views be said to be implied by quantum mechanics?

I was glad to see that Professor Wise, an expert on late-nineteenth-century physics, finds no error in what I had to say about the history of science. Unfortunately he does find a great many errors in things that I did not say. I never said there were no physicists in the early twentieth century who found cultural or philosophical implications in relativity or quantum mechanics, only that in my view these inferences were not valid. I never said that the apparent subjectivism of quantum mechanics was "of no great historical significance," only that I think we know better now. Just as anyone may get inspiration from scientific discoveries, scientists in their work may be inspired by virtually anything in their cultural background, but that does not make these cultural influences a permanent part of scientific theories. I never tried "to expunge all mystical physicists" as well as "creationists and other religious enthusiasts" from the history of science. I did say that I had never met a physicist who was a mystic, but my article had nothing to say about the frequency of other forms of religious belief among scientists, past or present.

On the subject of mystical physicists, it is interesting that when

Professor Wise tries to find up-to-date examples, he can get no closer than two physicists whose major work was done more than sixty years ago. He expresses surprise that no physicist has yet presented string theory as a form of Platonic mysticism, but I think I can explain this. It is because we expect that string theory *will* be testable—if not directly, by observing the string vibrations, then indirectly, by calculating whether string theory correctly accounts for all of the currently mysterious features of the standard model of elementary particles and general relativity. If it were not for this expectation, string theory would not be worth bothering with.

I tried in my article to put my finger on precisely what divides me and many other scientists from cultural and historical relativists by saying that the issue is not the belief in the reality of the laws of nature. Professor Wise makes a good point that, in judging the reality of the laws of nature, the test is not just their validity, but also their lack of "multiplicity." Indeed, as I wrote in my article, one of the things about laws of nature like Maxwell's equations that convinces me of their objective reality is the absence of a multiplicity of valid laws governing the same phenomena, with different laws of nature of different cultures.

(To be precise, I don't mean that there is no other valid way of looking at the electric and magnetic phenomena that Maxwell's equations describe, because there are mathematically equivalent ways of rewriting Maxwell's theory, and the theory itself can be replaced with a deeper theory, quantum electrodynamics, from which it can be derived. What I mean is that there is no valid alternative way of looking at the phenomena described by Maxwell's equations that does not have Maxwell's equations as a mathematical consequence.)

Whatever cultural influences went into the discovery of Maxwell's equations and other laws of nature have been refined away, like slag from ore. Maxwell's equations are now understood in the same way by everyone with a valid comprehension of electricity and magnetism. The cultural backgrounds of the scientists who discovered such theories have thus become irrelevant to the lessons that we should draw from the theories. Professor Wise and some others may be upset by such distinctions because they see them as a threat to their own "agenda," which is to emphasize the connections between scientific discoveries and their cultural context; but that is just the way the world is.

On the other hand, the gap between science and other forms of culture may be narrow or absent for the sciences that specifically deal with human affairs. This is one of the reasons that in writing of this gap in my article, I wrote about physics, and explicitly excluded science like psychology from my remarks. I concentrated on physics also because that is what I know best. Professors Holquist and Shulman are mistaken in thinking that when talking of "science," I meant just physics, and excluded "microbiology, genetics, and the new brain sciences." I was pretty careful in my article to write of physics when I meant physics, and of science when I meant science. I can't see why Professor Shulman, a distinguished molecular biophysicist and biochemist, should be unhappy with my *not* offering opinions about the cultural implications of biology.

I should perhaps have made more clear in my article that I have no quarrel with most historians, philosophers, and sociologists of science. I am a fan of the history of science, and in my recent books I have acknowledged debts to writings of numerous historians, philosophers, and sociologists of science.[29] In contrast with Alan Sokal, who in perpetrating his hoax was mostly concerned about a breakdown of the alliance between science and the political left, my concern was more with the corruption of history and sociology by postmodern and constructivist ideologies. Contrary to what Professor Levine may think, my opposition to these views is not due to any worry about the effect they may have on the economic pinch hurting science. In years of lobbying for federal support of scientific programs, I never heard anything remotely postmodern or constructivist from a member of Congress.

Among philosophers of science, Thomas Kuhn deserves special mention. He was a friend of mine whose writings I often found illuminating, but over the years I was occasionally a critic of his views.[30] Even in his celebrated early *Structure of Scientific Revolu-*

tions, Kuhn doubted that "changes of paradigm carry scientists and those who learn from them closer and closer to the truth." I corresponded with him after we met for the last time at a ceremony in Padua in 1992, and I found that his skepticism had become more radical. He sent me a copy of a 1991 lecture,[31] in which he had written that "it's hard to imagine . . . what the phrase 'closer to the truth' can mean"; and "I am not suggesting, let me emphasize, that there is a reality which science fails to get at. My point is rather that no sense can be made of the notion of reality as it has ordinarily functioned in philosophy of science." I don't think that it was "outrageous" for me to have said that, as I understood his views, Kuhn denied the objective nature of scientific knowledge.

Professor Levine and several others object to my criticism of Jacques Derrida, based as it seems to them on a single paragraph chosen by Sokal for mockery, which begins, "The Einsteinian constant is not a constant, is not a center. It is the very concept of variability—it is, finally, the concept of the game." When, in reading Sokal's *Social Text* article, I first encountered this paragraph, I was bothered not so much by the obscurity of Derrida's terms "center" and "game." I was willing to suppose that these were terms of art, defined elsewhere by Derrida. What bothered me was his phrase "the Einsteinian constant," which I had never met in my work as a physicist. True, there is something called Newton's constant which appears in Einstein's theory of gravitation, and I would not object if Derrida wanted to call it "the Einsteinian constant," but this constant is just a number (0.00000006673 in conventional units); I did not see how it could be the "center" of anything, much less the concept of a game.

So I turned for enlightenment to the talk by Derrida from which Sokal took this paragraph. In it, Derrida explains the word "center" as follows: "Nevertheless, . . . structure—or rather, the structurability of structure—although it has always been involved, has always been neutralized or reduced, and this by a process of giving it a center or referring it to a point of presence, a fixed origin."[32] This was not much help.

Lest the reader think that I am quoting out of context or perhaps just being obtuse, I will point out that, in the discussion following Derrida's lecture, the first question was by Jean Hyppolite, professor at the College de France, who, after having sat through Derrida's talk, had to ask Derrida to explain what he meant by a "center." The paragraph quoted by Sokal was Derrida's answer. It was Hyppolite who introduced "the Einsteinian constant" into the discussion, but while poor Hyppolite was willing to admit that he did not understand what Derrida meant by a center, Derrida just started talking about the Einsteinian constant, without letting on that (as seems evident) he had no idea of what Hyppolite was talking about. It seems to me that Derrida in context is even worse than Derrida out of context.

NOTES

1. Alan D. Sokal, "Transgressing the Boundaries: Toward a Transformative Hermeneutics of Quantum Gravity," *Social Text*, pt. I (spring/summer 1996): 217–52.

2. Alan D. Sokal, "A Physicist Experiments with Cultural Studies," *Lingua Franca*, pt. 2, "Revelation" (May/June 1996): 62–64.

3. In an afterword, "Transgressing the Boundaries," submitted to *Social Text*, Sokal explained that his goal was not so much to defend science as to defend the left from postmodernists, social constructivists, and other trendy leftists.

4. For instance, there is a linear relation between the number of calories in a cake and the amounts of each of the various ingredients: the graph of calories versus ounces of any one ingredient, holding the amounts of all the other ingredients fixed, is a straight line. In contrast, the relation between the diameter of a cake (of fixed height) and the amounts of its ingredients is not linear.

5. Operations are said to be noncommuting if the result when you perform several of them depends on the order in which they are performed. For instance, rotating your body by, say, thirty degrees around the vertical axis and then rotating it by thirty degrees around the north-south direction leaves you in a different position than these operations would if they were carried out in the opposite order. Try it and see.

6. Steve Fuller, letter to the *New York Times*, May 23, 1996, p. 28; and

Stanley Fish, "Professor Sokal's Bad Joke," Op-Ed article in the *New York Times*, May 21, 1996, p. 23.

7. Bruce Robbins and Andrew Ross, "Mystery Science Theater," *Lingua Franca*, pt. 2, "Response" (July/August 1996).

8. Robbins and Ross, "Mystery Science Theater."

9. Andrew Ross, "Introduction," *Social Text* (spring/summer 1996): 1–13.

10. Quoted by Robbins and Ross in "Mystery Science Theater."

11. Andrew Ross, *Strange Weather* (London: Verso, 1991), p. 42.

12. In general, the wave function of any system is a list of numbers, one number for every possible configuration of the system. For a single electron in an atom, the list includes a different number for each possible position of the electron. The values of these numbers give a complete description of the state of the system at any moment. One complication is that the possible configurations of any system can be described in different ways; for instance, an electron could be described in terms of its possible velocities, rather than its possible positions (but not by both at the same time). There are well-understood rules for calculating the numbers making up the wave function in one description if we know what these numbers are in any other description. Another complication is that these numbers are complex, in the sense that they generally involve the quantity known as i, equal to the square root of minus one, as well as ordinary real numbers.

13. See Jeremy Bernstein, *Hitler's Uranium Club* (Woodbury, N.Y.: American Institute of Physics, 1995).

14. For quotes and comments, see Jean Bricmont, "Science of Chaos or Chaos in Science?" *Physicalia Magazine* 17 (1995): 159–208, reprinted in *The Flight from Science and Reason* (New York: New York Academy of Sciences, 1996). A rejoinder and response are given by Ilya Prigogine and I. Antoniou, "Science of Chaos or Chaos in Science: A Rearguard Battle," *Physicalia Magazine* 17 (1995): 213–18; Jean Bricmont, "The Last Word from the Rearguard," *Physicalia Magazine* 17 (1995): 219–21.

15. Nonlinear dynamics deals with cases in which the rates of change of various quantities depend nonlinearly on these quantities. For instance, the rates of change of the pressures, temperatures, and velocities at various points in a fluid like the atmosphere depend nonlinearly on these pressures, temperatures, and velocities. It has been known for almost a century that the long-term behavior of such systems often exhibits chaos, an exquisite sensitivity to the initial condition of the

system. (The classic example is the way that the flapping of a butterfly's wings can change the weather weeks later throughout the world.) For physicists, the current interest in nonlinear dynamical systems stems from the discovery of general features of chaotic behavior that can be precisely predicted.

16. Bruce Robbins and Andrew Ross, letter to the *New York Times*, May 23, 1996, p. 28.

17. *Dreams of a Final Theory* (New York: Pantheon, 1993).

18. Sandra Harding, *The Science Question in Feminism* (Ithaca, N.Y.: Cornell University Press, 1986), pp. 9, 250.

19. Fish, "Professor Sokal's Bad Joke."

20. Andrew Ross was quoted by the *New York Times* on May 18, 1996, to the effect that "scientific knowledge is affected by social and cultural conditions and is not a version of some universal truth that is the same in all times and places."

21. Bruno Latour, *Science in Action* (Cambridge: Harvard University Press, 1987).

22. For instance, see Thomas Kuhn, "The Road Since Structure," in *PSA 1990* (Philosophy of Science Association, 1991), and "The Trouble with the Historical Philosophy of Science" (1991 lecture published by the Department of the History of Science, Harvard University, 1992).

23. Sarah Franklin, "Making Transparencies—Seeing through the Science Wars:" *Social Text* (spring/summer 1996): 141–55.

24. This point of view was expressed to me by the historian Harry Collins, then at the Science Studies Centre of the University of Bath.

25. In "Independence, Not Transcendence, for the Historian of Science," *ISIS* (March 1991), Paul Forman called for historians to exercise an independent judgment not just of how scientific progress is made, but even of what constitutes scientific progress.

26. See especially Gerald Holton, *Science and Anti-Science* (Cambridge: Harvard University Press, 1993), and Paul R. Gross and Norman Levitt, *Higher Superstition* (Baltimore: Johns Hopkins Press, 1994). The issue of *Social Text* in which Sokal's hoax appeared was intended as a response to Gross and Levitt's book, which also, according to Sokal, inspired his hoax.

27. On this, see Herbert Butterfield in *The Origins of Modern Science* (New York: Free Press, 1957), especially chap. 2.

28. Gerald Holton, *Einstein, History, and Other Passions* (Reading, Mass.: Addison-Wesley, 1996), p. 129.

29. This includes contemporary historians of science like Laurie Brown, Stephen Brush, Gerald Holton, Arthur Miller, Abraham Pais, and Sam Schweber; sociologists of science like Robert Merton, Sharon Traweek, and Stephen Woolgar; and philosophers like Mario Bunge, George Gale, Ernest Nagel, Robert Nozick, Karl Popper, Hilary Putnam, and W. V. Quine. These references can be found in Steven Weinberg, *Dreams of a Final Theory* (New York: Pantheon, 1993), and *The Quantum Theory of Fields* (Cambridge: Cambridge University Press, 1995). There are many others whose works I have found illuminating, including the historian of science Peter Galison, the sociologist of science Harriet Zuckerman, and the philosophers Susan Haack and Bernard Williams.

30. See *Dreams of a Final Theory*, and "Night Thoughts of a Quantum Physicist," *Bulletin of the American Academy of Arts and Sciences* 69, no. 3 (December 1995): 51.

31. Thomas Kuhn, "The Trouble with the Historical Philosophy of Science," Rothschild Distinguished Lecture, November 19, 1991 (Department of the History of Science, Harvard College, 1992).

32. Jacques Derrida, "Structure, Sign, and Play in the Discourse of the Human Sciences" in *The Structuralist Controversy*, ed. R. Macksey and E. Donato (Baltimore: Johns Hopkins University Press, 1972), p. 247.

STUDY QUESTIONS

Much to Matt Cartmill's disgust, Donna Haraway says that she will treat science as a narrative and primatology as a sort of science fiction. Cartmill suspects, apparently with good reason, that Haraway intends to debunk science by regarding scientific accounts as mere stories and as no more "rational" or "objective" than other kinds of stories. However, do you necessarily debunk science by reading it as literature? What could we learn about science by reading it as literature, that is, examining its metaphors, narrative structure, rhetoric, and so on? Unlike a TV docudrama, science cannot be loosely "based on a true story." Still, could you tell a good story and still be doing good science?

Critics accuse postmodernists of obscurantism; that is, they are accused of expressing themselves in an unnecessarily opaque and verbose jargon that prevents a clear understanding of their claims and arguments. In your opinion is Haraway guilty of such obscurantism? Can you find passages where you think that she is being intentionally obscure or needlessly difficult? Postmodernists reply that "plain" speech or "standard" English is rife with assumed biases and so cannot be used without reinforcing such prejudices and distortions. Is there any substance to this claim? Is there ever an excuse, unless you want to be misunderstood, for writing less clearly than you can? Is objectivity possible in a field like primatology? Can we avoid reading our own cultural assumptions and political agendas into the study of creatures so akin to ourselves? Perhaps some hard data can be gotten and reliable observations made (e.g., "Male *A* attempted to mate with female *B* three times during a two-hour period"). Still, data have to be interpreted, and how do we keep our interpretations free of bias?

Much of the issue between Weinberg and his critics seems to resolve into a dispute about just where social influence is found in science. Weinberg admits that there are many kinds of cultural, religious, or ideological influences on scientists. For instance, many have been mystics or political activists. Still, he argues that when a scientist's work is taken up by the community and made a part of the structure of science, it is purged of its nonscientific elements and means the same for everybody. Though he does not mention it, Weinberg seems to be relying on the distinction that philosophers of science have made between the "context of discovery" and the "context of justification." That is, all sorts of nonscientific influences or motivations may lead a scientist to a discovery, but once the discovery is made and confirmed by the relevant scientific community, such associations are purged. Can this distinction be challenged, or is it an essential insight into the nature of science? That is, do some theories accepted by scientific communities still bear the nonscientific stamp of their origins, or does science succeed, over time perhaps, in rooting these out?

PART 4

CONSERVATIVE CRITIQUES OF SCIENCE

INTRODUCTION

Left-wing intellectuals incited much of the recent controversy over science. The social constructivist, feminist, and postmodernist science critics may all be regarded as representatives of the "academic left." However, conservative thinkers have also raised questions, sometimes profound ones, about the nature and practice of science, its role in society, and its significance for human life. In fact, the right has traditionally challenged science more often than the left. Conservative intellectuals have questioned science when it seemed to undermine the established social order. Most often, conservatives were stirred when they perceived science as inimical to religious traditions.

What is the relation between religion and science? At one time the "warfare" model predominated. Nineteenth-century biologist and polemicist T. H. Huxley expressed this theory when he said that "extinguished theologians lie around the cradle of a new science" like the poisonous snakes strangled by the infant Hercules. According to the "warfare" view, religion and science are inevitably opposed, and the one can advance only at the expense of the other. For a new science to develop, obscurantist theologians have to be (metaphorically!) strangled. Progressive-minded people will therefore regard science as the beacon of enlightenment and religion as the dustbin of dogma.

Historians of science no longer accept the "warfare" model. True, there have been (and are) conflicts between science and religion. Some religious people have unwisely attempted to block scientific progress when they perceived a threat to orthodoxy. Conversely, some atheist scientists have enjoyed tweaking the noses of believers. But the more usual relation between religion and science has been peaceful coexistence if not symbiosis. Many of the greatest scientists were deeply, if sometimes eccentrically, religious. Many theologians and religious leaders have been enthusiasts for science, or even scientists in their own right (Gregor Mendel, Teilhard de Chardin, and Georges Lemaître come to mind).

Still, some conservative religious thinkers hold that science, or at least some of its claims, practices, or presuppositions, are inimical to religion. It is no surprise that some of the sharpest controversies still concern Darwinism. As Daniel Dennett, a staunch Darwinian, puts it, Darwinism is "universal acid"; it threatens to corrode all ideologies. There is a strong temptation, not only among explicitly religious people, to block off certain aspects of the world and forbid them from getting the kind of explanation Darwinism offers. For instance, if Darwinism is true, the human mind is the product of a natural, historical process in which our distinctive (and precious) mental capacities—including our consciousness, creativity, moral sense, and religious yearnings—developed gradually from earlier, more primitive feelings and abilities. In short, brains and their functions have evolved. The most refined human sensibility grew out of animal passion; the highest intellectual capacity was honed in a brute struggle for survival.

For many people, this is just an unacceptable conclusion. For them it is intuitively clear that consciousness and our highest mental aptitudes are of a fundamentally distinct kind that cannot, in principle, be reached by any degree of evolutionary development from more primitive capacities. Even secular philosophers, like Thomas Nagel, have expressed doubts.* Nagel thinks that evolutionary theory can say why we have those capacities that aid us in the struggle to survive and pass on our genes—finding food, mates, shelter, and so on. But how can we explain those higher mental states and aptitudes such as consciousness or aesthetic appreciation? Prima facie the ability to understand quantum mechanics or Godel's theorem, which (some) people have, has no bearing on such very prosaic matters as getting food or avoiding predators. Even Alfred Russel Wallace, who discovered natural selection independently of Darwin, held that human mental abilities begged for a transcendent cause.

If features of the natural world thus seem to cry out for explanation in terms of the supernatural or transcendent, why are scientists so reluctant to invoke such causes? Some conservative thinkers say it is because of science's dogmatic commitment to naturalism. Science is natural science. Science in general (and Darwinism in particular) is committed to finding natural causes of observed phenomena. From archaeology to zoology, no current science will admit explanations in terms of gods, demons, angels, or other supernatural beings. For instance, a volcanologist who attempted to invoke the goddess Pele to account for the variations of lava flow from Kilauea would lose all standing in the field. However much we might respect native Hawaiians' belief in Pele (and we should), the science of geology just will not place goddesses alongside natural causes as potential explanations of phenomena.

According to law professor and antievolution activist Phillip E. Johnson, the dogmatic commitment to naturalism leads science to

*See Thomas Nagel, *The Last Word* (Oxford: Oxford University Press, 1997) for his argument.

propose wildly inadequate evolutionary explanations for phenomena that manifestly seem to be products of intelligent design. Scientists reply that their commitment is to methodological, not metaphysical, naturalism. That is, naturalism is a working hypothesis, or better, a background assumption that restricts attention to natural causes as a matter of methodological expedience. It is a heuristic principle, that is, an assumption that we think will guide us along fruitful paths of inquiry, not a categorical denial of the existence of the supernatural. Further, they point out, that assumption has served science very well. Guided by naturalistic principles, science has succeeded spectacularly in finding natural causes of many phenomena, some once considered so marvelous that our ancestors invoked the gods to account for them. Johnson will have none of this. He views methodological naturalism as a front for those who outright deny the existence of the supernatural. He quotes a number of prominent scientists' pronouncements denying the supernatural.

Religious conservatives are not the only ones who have seen naturalistic science as draining depth and beauty from the world. The romantic poets, hardly noted for their religious orthodoxy, often charged that science deprives the world of its awesome mystery, reducing nature's spiritual richness to a cut-and-dried collection of facts and figures. John Keats, one of the great poets of English literature, famously said, "All charms fly at the touch of cold philosophy" (science was called "natural philosophy" in those days). He declared that science could "clip an angel's wings" and "unweave the rainbow," that is, the power of nature to inspire or terrify (the romantics celebrated even the wild, uncontrolled aspects of nature) is abridged by a science that makes everything dull and commonplace.

A case in point is comets. Comets were once terrible portents in the heavens, signs of divine wrath and harbingers of plague, famine, or war. Comets "blaze forth the death of princes," said Shakespeare. But in the eighteenth century, astronomer Edmund Halley, using the principles of Newton's mechanics, showed that comets are predictable and orderly denizens of the solar system, blindly obedient to the laws of physics like planets and moons. More recent science has shown comets to be very unromantic objects—"dirty snowballs" as Carl Sagan called them. Surely, scientists will say, science should be credited for removing the mystery of comets and thereby ending the irrational terror they used to excite. The romantics would reply that this is precisely the point. A nature that can no longer terrify can also no longer inspire awe or overwhelm with its magnificence. Seeing nature through the eyes of science is like seeing the emperor naked; a figure that had been clothed in terrible majesty is revealed as pale, frail, and ordinary.

Of course, the charges of religious and romantic critics have not gone unanswered. Perhaps the fairest reply to advocates of intelligent design theories came from Darwin himself. In the final chapter of *Origin of Species*, Darwin considers hypotheses of divine design versus his hypothesis of natural selection. He does not invoke a metaphysical or even a methodological naturalism, but does his best to test the theory of special creation against his account. If organisms possess their adaptive features due to special creation, that is, if a benevolent deity designed them to fulfill their role in the world, what should we find in nature? What we do find are all sorts of natural features that are either inexplicable on or contrary to theses of benevolent design, but quite understandable given the blind, amoral, opportunistic process of natural selection.

Why would God create two groups of organisms, the sharks and the rays, for instance, which are so much alike that with some species it is hard to tell where the sharks end and the rays begin? The mystery becomes explicable if we see sharks and rays as diverging from a common ancestor who had features shared by both groups. Why would the pampas of Argentina have woodpeckers when there is no tree within hundreds of miles? Surely, a thoughtful designer would have populated grasslands with creatures specifically designed to live there. However, if the pampas woodpecker adapted through natural selection as its environment gradually changed from forest to grasslands, the mystery is solved. Then there are those hateful creatures, like pathogens and parasites, who make a living by causing the suffering of other organisms. Darwin's favorite example, when he played the role of "devil's chaplain," was the wasps of the Ichneumon family. These creatures lay their eggs on the paralyzed, but still living, bodies of caterpillars. When the eggs hatch, the wasp larvae, with infinite patience and skill, slowly devour the caterpillar, keeping it alive as long as possible by leaving the most vital organs to the last. Surely, a benevolent deity would have devised something other than bloodsucking parasites, the rabies, HIV, and Ebola viruses, and the ghastly Ichneumon wasps. But natural selection favors whatever works, whether we regard it as wasteful, disgusting, or horrifying. Thus, it seems that there is much that special creation cannot explain and evolution by natural selection can.

Of course, defenders of special creation could say that God has reasons for making things the way he did and that we humans just cannot know those reasons. But this reply misses the point. The point is that a scientific hypothesis is supposed to explain things. Suppose that, as many philosophers argue, even the grossest evils that occur cannot be proven incompatible with God's existence. The fact remains that the natural world contains many troubling features that special creation cannot explain but Darwinian evolution can.

What about the charge that science turns nature into a desiccated catalogue of facts, and so deprives it of its mystery and wonder? It is possible to take a wholly scientific view, yet to see the universe as charged with mystery, beauty, and "sacred depths." Many scientists view the idea of the romantic poets that science can "unweave the rainbow," as perversely and perniciously false. They view the truth as precisely opposite, that is, that the more we know about the universe, the more it evokes our awe.

Aristotle, the first biologist, stated once and for all the wonder that can be found even in the meanest creature:

> So we must not approach the examination of the lower animals with a childish distaste; in all natural things there is something wonderful. There is a story about some people who wanted to meet Heraclitus. When they came to visit him they saw him in the

> kitchen, warming himself at the stove, and they hesitated. But Heraclitus said, "Don't be afraid to come in. There are gods even here." In the same way, we ought to approach our enquiries about each kind of animal without any aversion, knowing that in every one of them there is something natural and beautiful. For in the works of nature orderliness and purpose are to be found in the highest degree, and the end for the sake of which they have been composed and have come into being is itself a kind of beauty (Parts of Animals I.5.645a5).

The same sense of awe is expressed by another biologist of some repute:

> . . . from the war of nature, from famine and death, the most exalted object which we are capable of conceiving, namely, the production of the higher animals, directly follows. There is a grandeur in this view of life, with its several powers, having been originally breathed by the Creator into a few forms or into one; and that, while this planet has gone circling on according to the fixed law of gravity, from so simple a beginning endless forms most beautiful, and most wonderful, have been and are being evolved.

This, of course, is from the final paragraph of *Origin of Species*. Time and again, the most tough-minded scientists can be found expressing their wonder and delight at the world their science discloses. The debate over these issues is ancient, but far from over.

The readings for this section begin with the chapter "The Grand Metaphysical Story of Science" from Phillip E. Johnson's book *Reason in the Balance*. Johnson considers the quest of scientists like Stephen Hawking for a theory of everything (TOE). A TOE would provide a mathematical formalism that would give us a unified account of the most basic forces and interactions of nature. Johnson argues that there are no reasonable prospects for getting a TOE that would be testable. The theory would be too mathematically complex and any tests that could in principle be performed would require resources that the whole earth could not generate.

Speculations about the possible shape of such TOEs, such as the one Hawking proposes in *A Brief History of Time*, are precisely that, says Johnson—speculations. They have no authority other than the prestige of those who announce them. In fact, says Johnson, the yearning for a TOE is the product of a metaphysical prejudice on the part of many scientists—the bias against allowing any room for a creator and the insistence that naturalistic explanations be the only kind. This dogmatic naturalism, says Johnson, is assumed in other fields such as neuroscience. Neuroscientists attempt to explain thought and consciousness in terms of the function of the brain; spiritual entities, such as souls, are ruled out. Johnson charges that this effort will be self-defeating because it will imply that human thinking is merely an accidental by-product of natural law rather than a rational process.

Robert T. Pennock replies to Johnson and other creationists in his book *Tower of Babel*. In a selection from that book, Pennock argues that scientists' naturalism is not a metaphysical bias but a legitimate methodological requirement. Pennock argues that the primary commitment of science is not to any particular metaphysical position, but to a method of inquiry. The rigorous testing of theory against data, which is the heart of the scientific enterprise, requires that the theories we seek to evaluate be testable. However, theories postulating supernatural causes are notoriously difficult to test. For instance (this is my example, not Pennock's), on very rare occasions patients suffering from advanced metastatic cancer experience a complete, spontaneous remission of their disease. Medical science does not know why this happens. Suppose someone were to propose that in such cases God has intervened with a healing miracle. How could this hypothesis be confirmed? What kind of practical test, experiment, or observation could possibly confirm that such healings are miraculous rather than due to presently unknown natural causes? Pennock also points out that supernatural hypotheses would also provide too easy an explanation in many circumstances. To say "God does it" when confronted with something hard to understand provides a too-convenient rug under which we can sweep our ignorance. Finally, Pennock examines Johnson's proposal that natural science be replaced with theistic science. Pennock notes that this proposal raises the question of what additional sources of knowledge, in addition to naturalistic methodologies, such "theistic science" could employ. Revelation? The design argument? Pennock argues that no such alternatives provide knowledge worthy of the name "science."

Finally, the romantic objection to "soulless" science and the scientific response are presented in Martin Gardner's article "Science vs. Beauty?" For over fifty years Martin Gardner has explained and defended mathematics and science in numerous books and articles. His essay provides a generous sampling of literary critiques of science and the replies by Richard Feynmann and other well-known scientists.

ELEVEN

THE GRAND METAPHYSICAL STORY OF SCIENCE

PHILLIP E. JOHNSON

Stephen Hawking's book *A Brief History of Time* was one of the greatest surprise successes of publishing history, having sold an estimated five million copies worldwide by 1992, after an initial printing of only five thousand in 1988. This level of success had to be due to something other than the scientific content of the book. Very few of the millions of purchasers could have understood Hawking's descriptions of relativity theory and quantum mechanics, which are difficult for nonphysicists to follow even in popularized form. In any case, essentially the same information was available in many other popular books about physics and cosmology.

What Hawking's book offered that the others did not was Hawking himself, a man who had triumphed over a horribly crippling disease to reach the pinnacle of scientific fame. It also offered one of the most ambitious programs for the scientific enterprise that has ever been stated.

THEORIES OF EVERYTHING

In Hawking's expansive vision, theoretical physics promises to provide humanity with something more than mathematical theories understandable only to specialists, and something much more important to ordinary people than possible technological benefits like new sources of energy. What Hawking envisions is a kind of universal human wisdom attainable under the tutelage of physicists. The goal is nothing less than "a complete understanding of the events around us, and of our own existence."

The first big step toward this goal is a unified theory of the four fundamental forces of nature, the final theory that particle physicists dream of. Once the unified theory is discovered, Hawking thinks, it will be possible to teach philosophers and even ordinary people something of what the theory means. Eventually this knowledge will enable everyone to take part in a great conversation about such grand questions as why it is that we and the universe exist. Finding the answer to the riddle of existence, Hawking concludes, "would be the ultimate triumph of human reason—for then we would know the mind of God."

To have a complete understanding of everything that happens, and even why the universe exists at all, can certainly be described as knowing the mind of God—or more precisely, attaining the omniscience previously attributed to God. Perhaps it is not surprising that a book captures the public imagination when it holds out even a distant prospect of omniscience to the world at large and seems to back the promise with the authority of science. But how much can scientific investigation really contribute to an understanding of the meaning of our existence, particularly a science that is as incomprehensible to nonspecialists as theoretical physics?

The scientific side of the Hawking story begins with the work of another great mathematical physicist, Roger Penrose, famed as the discoverer of "black holes." Penrose demonstrated that the existence of these exotic astronomical entities follows as a matter of logic from the gravitational principles of Einstein's general theory of relativity. According to Hawking's account, Penrose

> showed that a star collapsing under its own gravity is trapped in a region whose surface eventually sinks to zero size. And, since the surface of the region sinks to zero, so must its size. All the matter in the star will be compressed into a region of zero volume, so the density of matter and the curvature of space-time become infinite. In other words, one has a singularity contained within a region of space-time known as a black hole.

Hawking's own scientific fame began with a 1970 paper, coauthored with Penrose, which applied the black hole theory to the universe as a whole. Just as a black hole is a dying giant star collapsing into a point of zero volume, the universe is now thought to be expanding from a similar infinitesimal point. As the black hole ends in a singularity, the universe must have begun with one. The existence of a singularity at the beginning of time had previously been thought to be a possible implication of general relativity; Hawking and Penrose demonstrated that it was an inevitable implication.

This conclusion of theoretical physics has great philosophical and religious significance. The big bang theory itself had to overcome considerable resistance from scientists: they objected to the implication that the universe had a definite beginning in time, because such a moment of creation was previously associated more with religion than with science. Indeed, the proposition that the universe began in an unimaginably vast explosion of energy sounds to many like a rendering in modern scientific language of what the Book of Genesis said long ago: "God said, Let there be light!"

Despite these metaphysical objections, the big bang theory triumphed because it fit the evidence better than any rival. An expanding universe seems to follow from the logic of general relativity, the expansion seems to be confirmed by observations of the Hubble "red shift," and the primeval explosion seems to have left an "echo" in the form of a universal background radiation that was first detected in 1965.

Still, Hawking and Penrose's demonstration that the universe must have begun with a singularity significantly increased the metaphysical uneasiness. A singularity is defined in relativity jargon as a point at which the space-time curvature becomes infinite, which sounds innocent enough, but the disturbing implication is that at such a point all relativity-based laws of nature break down. The existence of a singularity at the absolute beginning thus could be taken to imply that there was a time when the most fundamental laws of science did not exist. How, then, did the laws come into existence? Whatever the scientists might prefer to think, it was inevitable that the popular culture would infer that God—a supernatural entity unknown to science—must have created the laws.

That kind of answer is intensely disliked by many scientists, including Hawking, because to leave any room for the supernatural seems to leave science incomplete. The search was on for some way to discard the singularity. The search involves the attempt to provide a grand unified theory (GUT) of physics, a project that has been celebrated in so many books and educational television programs that most readers have undoubtedly heard of it.

Physicists recognize four fundamental forces. Three of these—electromagnetism, the weak nuclear force, and the strong nuclear force—operate at subatomic levels and are understood through the mathematics of quantum theory. The fourth, gravity, is important for large-scale bodies and is understood in terms of Einstein's general theory of relativity. In our world the forces are distinct, but many physicists think that they were unified at the earliest moment of the big bang, when the universe was unimaginably hot and dense.

A satisfactory "electroweak" theory unifying electromagnetism and the weak nuclear force already exists, and it has achieved some experimental confirmation. Although the strong nuclear force has its own theory, called quantum chromodynamics, physicists think they have made some progress toward a unification of quantum chromodynamics with the electroweak theory. The great difficulty lies in extending the program of unification to the fourth force, gravity, because the principles of relativity and quantum mechanics are contradictory.

Whether a unified theory can ever be achieved is in dispute. One possibly insuperable obstacle is the extreme difficulty of carrying out the experiments needed to test the work of theoreticians. According to David Lindley's *The End of Physics: The Myth of a Unified Theory,* particle physics is becoming more like myth than science, because its mathematical constructions are so distant from any conceivable experimental confirmation. If the Texas Supercollider had been built, it might have found the famous Higgs particle, which would have provided further support for the electroweak theory—and thus for the proposition that the physicists are in general on the right track. The Supercollider has been canceled, however, and in any case the experiments needed to pursue the unification project to the next stages would require energy levels that cannot be produced on earth, especially at any cost that the public is likely to be willing to fund.

In consequence a proposed final theory could not be confirmed by experiment, but only by agreement among the theoreticians, which makes the theory sound more like philosophy than science. But the inspiration provided by past success overcomes such practical obstacles in the minds of theorists, and so popular books by Hawking and other leading physicists imply confidence that the grand unification project can be carried out—possibly even while the authors are still alive to enjoy the triumph.

The extraordinary metaphysical significance of a unified theory of the four forces is illustrated by the names physicists give to it:

"the holy grail of physics," "the final theory," and especially "the theory of everything." These romantic titles are not justified by any tangible accomplishments expected to flow directly from the theory itself, since even a successful unified theory is not expected to tell us anything in particular about such mysteries as how galaxies and galactic clusters form, how life arose or why people behave as they do. The immense imaginative appeal of a unified theory stems entirely from the position it occupies in the naturalistic philosophy that scientists generally assume in their work.

If nature is really a permanently closed system of physical causes and effects, then everything that has happened in the entire history of the cosmos must be determined (or at least permitted) by the conditions that existed at the beginning. If in the beginning nothing existed except the laws and the particles, and nothing fundamentally new has entered the universe subsequently, then a complete understanding of conditions at the beginning is in principle the key to a complete understanding of everything that followed. The unified theory is therefore what might be called the opening chapter in the grand metaphysical story of science, and the set of laws described by the unified theory is the scientific equivalent of a creator. That is why religious language permeates the books about the theory, and why Hawking thinks that to achieve a complete understanding of the theory would be to know the mind of God—in the sense of knowing all that there is to know.

To be truly like God, though, the theory has to be eternal, and the existence of a singularity at the beginning implies a time when the laws described by the theory did not exist. Hawking's greatest contribution to what might be called the religious unification of physics was to propose a theory that gets rid of the awkward singularity whose existence he and Penrose had established in the first place. What he suggested was that the big bang singularity might merely imply that at the earliest moment the gravitational field was so strong that general relativity was not applicable, and scientists may be able to employ a "quantum theory of gravity" (if one is ever discovered) to understand the beginning. A quantum theory of gravity would probably involve "imaginary time," a mathematical concept employed to resolve certain problems in quantum mechanics. In imaginary time, explains Hawking, "the distinction between time and space disappears completely." Accordingly, a theorist can do away with the embarrassing singularity at the beginning by doing away with the need for a beginning. Here is how Hawking puts it:

> The quantum theory of gravity has opened up a new possibility, in which there would be no boundary to space-time and so there would be no need to specify the behavior at the boundary. There would be no singularities at which the laws of science broke down and no edge of space-time at which one would have to appeal to God or some new law to set the boundary conditions for space-time. One could say: "The boundary condition of the universe is that it has no boundary." The universe would be completely self-contained and not affected by anything outside itself. It would neither be created nor destroyed. It would just BE.

Hawking freely concedes that the no-boundary concept of space-time is "just a *proposal:* it cannot be deduced from some other principle." He goes on to say that such a proposal may be put forward for aesthetic or metaphysical reasons, but the test of its status as a scientific theory is whether it makes predictions that can be verified by observation.

Obviously, the no-boundary proposal cannot make such predictions at the present time. That is not only because the necessary unification of relativity and quantum theory has not been achieved, but also because any model that described the whole universe would be too complicated mathematically to permit exact predictions. Theorists therefore would have to make simplifying assumptions and approximations, and Hawking concedes that "even then, the problem of extracting predictions remains a formidable one." One could never know whether any hypothetical predictive success was due to the particular simplifying assumptions that the theorist chose to make. David Lindley seems to be right; without experimental confirmation, theoretical physics takes on the qualities of myth.

The no-boundary proposal is therefore properly labeled as an element in a grand metaphysical story rather than as a scientific theory. When metaphysical stories are told by world-famous physicists, however, they take on an air of authority and may have profound consequences on the thinking of persons who do not distinguish between "stories scientists tell" and "scientific theories."

Paul Davies, another prominent mathematical physicist, explains the importance of Hawking's no-boundary proposal in his book *The Mind of God,* in a chapter titled "Can the Universe Create Itself?" Davies candidly states at the outset, "This particular explanation may be quite wrong," but immediately adds that the correctness of any particular explanation does not matter. According to Davies, "What is at issue is whether or not some sort of supernatural act is necessary to start the universe off. If a plausible scientific theory can be constructed that will explain the origin of the entire physical universe, then at least we know a scientific explanation is possible, whether or not the current theory is right."

But an untestable metaphysical story that makes no predictions is not a scientific theory, and such speculation cannot prove that a true naturalistic explanation for the ultimate beginning exists. Davies's statement amounted to saying that even a myth is sufficient to reassure scientific naturalists that they do not have to worry about the supernatural.

Scientists and philosophers of science frequently say that God is a subject outside of science, but such statements are seriously misleading. It would be more accurate to say that the scientists who think about the big picture are obsessed with the God issue, and it is natural that they should be. The aim of historical scientists—those who attempt to trace cosmic history from the big bang or before to the present—is to provide a complete naturalistic picture of reality. This enterprise is defined by its determination to push God out of reality, because naturalism is defined by its exclusion of the supernatural. Particle physicists and cosmologists tend to be very religious people in their own way, but their religion is often science itself, and so the only creation story they will accept is one in which all the elements of reality are in principle accessible to scientific investigation. An imaginative story that makes the universe itself eternal is hence preferable to a scientific theory that requires a disturbing singularity at the beginning, and for this reason the former may attain the status of scientific knowledge on its imaginative appeal alone.

That removing God from the history of the cosmos is the central point of *A Brief History of Time* is pointed out to readers by the astronomer Carl Sagan, in the closing lines of his introduction to the book, although Sagan presents the conclusion as if it were an unanticipated experimental result rather than the conscious purpose of the author:

> The word God fills these pages. Hawking embarks on a quest to answer Einstein's famous question about whether God had any choice in creating the universe. Hawking is attempting, as he explicitly states, to understand the mind of God. And this makes all the more unexpected the conclusion of the effort, at least so far: a universe with no edge in time, no beginning or end in time, and nothing for a Creator to do.

MIND AND MATTER

One element in the imaginative dimension of *A Brief History of Time* consists of its telling a story about the ultimate beginning of the universe in a way intended to relieve scientific naturalists of their fear of the supernatural, while holding out to the public the prospect of eventual omniscience. A second element concerns Hawking's personal story, particularly visible in the publicity surrounding the book—and especially in the outstanding BBC television movie that was made from and about Hawking's phenomenal best-seller.

When as a young man Stephen Hawking was diagnosed with what Americans call Lou Gehrig's disease, he had every reason to give up on life. Instead, not only did he live much longer than the doctors had predicted, but he married, had children, and achieved preeminence in science. Today he occupies the Lucasian Professorship of Mathematics at Cambridge, a chair that once belonged to Sir Isaac Newton. That Hawking's publisher considers Hawking himself to rank with the greatest of scientists is suggested none too subtly to readers of *A Brief History of Time* by the inclusion of brief biographies of Einstein, Galileo, and Newton at the end of the slender volume.

When we see Hawking in person in the television adaptation of his book, his disease has left him almost totally helpless, unable to move from his wheelchair or even to speak normally. He speaks through a voice synthesizer by entering words into a computer program, producing an effect rather like a modern incarnation of the oracle of Delphi, or perhaps the supercomputer called Hal in the movie *2001: A Space Odyssey*. The thought that such a ruined body might hold a mind capable of penetrating the ultimate secrets of the universe is genuinely inspiring. In this sense Hawking's life is an archetype of the entire saga of science, for science is the story of the power of the mind to penetrate the fog of superstition and ignorance to discern the invisible reality beyond. The story of the man Stephen Hawking, in the mythological dimension that has so much to do with his immense popular success, is the story of mind over matter.

The irony is that what this heroic mind ends up producing is a reductionist science that reduces the mind itself to a trivial sideshow in a materialist universe. Hawking does not address this problem directly, but it surfaces in his book, for he recognizes that a physical theory of everything is inherentJy self-referential and hence potentially incoherent. The enterprise of science assumes that human beings—or scientists, at any rate—are rational beings who can observe nature accurately and employ logical reasoning to understand the reality behind the appearances. If a theory of everything exists, however, the laws it describes determine even the thoughts and actions of the scientists who aim to discover the theory. How then, wonders Hawking, can the scientists trust their own powers of reasoning? How can they know that the laws of physics predict or permit the discovery of a true theory?

Naturalistic philosophy offers one line of escape from this conundrum, and Hawking takes it. The only validation of the mind's reasoning power that science can provide is Darwin's principle of natural selection, which explains all adaptive features of organisms in terms of reproductive success. The theory posits that evolution rewarded those organisms that were best at drawing correct conclusions about the world and acting accordingly to escape predators, find mates, and so on. Right-thinking organisms would presumably excel at surviving and reproducing, and hence would leave more offspring than competitors who were more inclined to err. Eventually the ability to come to correct conclusions would become widespread in every population. In Hawking's words, "Provided the universe has evolved in a regular way, we might expect that the reasoning abilities that natural selection has given us would be valid also in our search for a complete unified theory, and so would not lead us to the wrong conclusions."

But one cannot avoid the problem of self-reference by invoking another theory in this way. Darwin's theory is just another product of the human mind, whose reasoning is still governed by the hypothetical theory of everything, so the problem of reliability is merely displaced rather than solved. In any case, Darwinian selection rewards only success in leaving offspring, and the presumption that abstract mental powers cause their possessor to leave more viable offspring than creatures who are more modestly endowed is neither borne out by experience nor even remotely plausible. By Darwinian criteria, the brains of rats and cockroaches are every bit as conducive to reproductive success as human brains, particularly when one reflects that the "fitness" of the human brain would have to be judged in the primitive conditions in which it supposedly evolved. One has only to consider Hawking's own hereditary disease to appreciate how little advanced mathematical gifts have to do with the ability to leave viable offspring.

There are ways of meeting this kind of objection, of course. An ingenious Darwinist can always suppose that the capacity to solve equations involving imaginary time evolved as a serendipitous byproduct of more modest mental traits that did increase reproductive success. The problem with this sort of explanation is not that it can be proved wrong, but rather that it is vacuous because it can "explain" any trait whatsoever. One might as well speculate—without any evidentiary support whatever—that the ability to solve equations is a secondary effect of a gene that also codes for a high sperm count or a pleasantly shaped nose.

Although invoking Darwinian selection does not solve the problem of self-reference, it is to Hawking's credit that he brought the problem to the attention of his readers, because what it really demonstrates is that a theory that is the product of a mind can never adequately explain the mind that produced the theory. The story of the great scientific mind that discovers absolute truth is satisfying only so long as we accept the mind itself as a given. Once we try to explain the mind as a product of its own discoveries, we are in a hall of mirrors with no exit. It is as if a disciple of Sigmund Freud were to explain his master's theory of the Oedipus complex as a product of Freud's own unconscious wish to murder his father and marry his mother. The theory could still conceivably be true of men in general, because Freud might just happen to be like others in this peculiar respect, but we would certainly not rely on Freud's authority to establish the point.

Attributing an idea to irrational unconscious desires or physical forces has about the same effect as showing that a judge received a huge cash payment from a litigant. The fact does not absolutely prove that the verdict was against the law or the evidence, because the litigant might conceivably have bribed the judge to do the right thing, but it mightily suggests the likelihood.

MATERIALIST THEORIES OF THE MIND

It is in the nature of explanation that one thing is explained in terms of something else that is assumed valid, and to explain the latter as nothing more than a product of the former is to create a logical circle. Yet naturalistic metaphysics is so seductive that eminent scientists and philosophers frequently do employ their own minds to attempt to prove that the mind is "nothing but" a product of physical forces and chemical reactions.

One of these is Francis Crick, the biochemist who as codiscoverer of the structure of DNA is almost as famous as Hawking himself. In his later years Crick has been drawn to the problem of consciousness, and he expressed his thoughts in the 1994 book *The Astonishing Hypothesis*. Here is how Crick states his own starting point:

> The Astonishing Hypothesis is that "You," your joys and your sorrows, your memories and your ambitions, your sense of personal identity and free will, are in fact no more than the behavior of a vast assembly of nerve cells and their associated molecules. . . . The hypothesis is so alien to the ideas of most people alive today that it can truly be called astonishing.

Of course the hypothesis is not astonishing at all to anyone acquainted with the recent history of science, because neuroscientists in particular have long taken for granted that the mind is no more than a product of brain chemistry. As Crick says, what makes the hypothesis astonishing is that it conflicts with the commonsense picture of reality most people assume as they go about the business of making decisions, falling in love, or even writing books advocating materialist reductionism.

The conflict with common sense would become apparent if Crick had presented his hypothesis in the first-person singular. Imagine the reaction of his publisher if Crick had proposed to begin his book by announcing that "I, Francis Crick, my opinions and my science, and even the thoughts expressed in this book, consist of nothing more than the behavior of a vast assembly of nerve cells and their associated molecules." Few browsers would be likely to read further. The plausibility of materialistic determinism requires that an implicit exception be made for the theorist.

Whatever common sense may have to say about the matter, the deconstruction of the mind advocated by Crick is implicit in the metaphysical materialism and naturalism that dominate the scientific community. Most biologists who express opinions on the subject in public take for granted that living organisms contain no "vital force" or other nonmaterial component, that complex organisms evolved from simpler predecessors by Darwinian selection, and that the human mind is therefore a product of material forces that valued nothing but success in reproduction. Given that understanding of things, what could the mind and its thoughts conceivably be but a product of the biochemistry of the brain, whatever the unlearned might think?

Crick does not claim that the materialistic understanding of the mind has been proved (the scientific part of his book limits itself to stating some tentative proposals for research into the mechanisms of vision). What he does claim is that the materialistic theory of the mind is the only possibility worth taking seriously. The only alternative he can envisage is some hangover from religion, which he characterizes contemptuously as "the superstitions of our ancestors."* Although Crick insists that scientists hold their hypotheses only as provisional beliefs and not by "blind faith," it is not clear what, if anything, could convince him that there is more to the mind than matter. Materialism to Crick is equivalent to science, and science to rationality.

Crick's reductionism is no idiosyncrasy; the same prejudice dominates contemporary science, regardless of reservations that individual scientists (who also have to live in the commonsense world) might express in private. It is not that workaday scientists are necessarily as enthusiastic about materialism as Crick is. The problem is that they do not know how to challenge materialism in principle without seeming foolish or sentimental. The materialists are intimidating because they seem to have the logic of science on their side. The same materialists are frustrated, however, because so many people are perversely unwilling to accept conclusions that a reductionist science necessarily implies. As the famous Stanford

---

*Crick's village-atheist level of understanding of religion is illustrated by the following comment: "Not only do the beliefs of most popular religions contradict each other but, by scientific standards, they are based on evidence so flimsy that only an act of blind faith can make them acceptable. If the members of a church really believe in a life after death, why do they not conduct sound experiments to establish it?" Religious people may take some comfort from the fact that Crick is nearly as contemptuous of philosophers: "Philosophers have had such a poor record over the last two thousand years that they would do better to show a certain modesty rather than the lofty superiority they usually display."

biochemist Arthur Kornberg complained to a 1987 meeting of the American Academy for the Advancement of Science, it is astonishing "that otherwise intelligent and informed people, including physicians, are reluctant to believe that mind, as part of life, *is* matter and *only* matter." On Kornberg's own premises, however, his astonishment was unjustified. Presumably, one kind of chemical reaction in the brain causes Kornberg to accept materialist reductionism, while another kind of reaction causes those physicians to doubt it.

There is a great deal at stake in the argument about whether the mind can really be explained as a strictly material phenomenon. The authority of the scientific priesthood rests on public acquiescence in the grand metaphysical story of science, but the public is manifestly inclined to doubt. In this delicate situation the rulers of science cannot afford to leave any openings for rival stories. They do not have to be able to supply a reductionist explanation of the mind today, or even tomorrow, but they do have to claim that their methods, and no others, are based on a correct understanding of what the mind really is. If they were to concede even tacitly that mental activity has its ultimate roots in something beyond physics and chemistry, the resulting opening for the supernatural would be far larger and more dangerous than that involved in a singularity at the beginning of time.

If science cannot explain consciousness, the way is open for some rival discipline—religion, in particular—to fill the vacuum with a different metaphysical story of great emotional or imaginative appeal. This explanation would be a better candidate for the title of "theory of everything" than anything particle physics or evolutionary biology can provide, because science itself is a product of the mind. Whoever explains the mind explains science, and gains authority to say how great or small a role science should play in the life of the mind. That is not an authority that scientists will voluntarily surrender to philosophers or theologians.

EVALUATING THE STORY OF SCIENCE

How are we to evaluate the grand metaphysical story of science? It would be a mistake to take the easy way out and dismiss the story because some of its key elements are not proved. Of course the elements are not proved; Hawking and Crick themselves are emphatic on that point with respect to the no-boundary proposal, the unified physical theory, and the materialist theory of the mind. The grand metaphysical story is not itself even conceivably subject to proof. Rather, it encapsulates the scientific way of thinking about the work that science has yet to accomplish.

The point of the materialist theory of the mind, for example, is not that such a theory exists (except in the most primitive and speculative form). It is that biochemists who are materialist reductionists fiercely want to believe that real progress toward understanding the mind comes only from learning the principles of biochemistry and not from listening to priests or philosophers. Similarly, the point of assuming that supernatural action played no role in the history of the cosmos is to inspire scientists with faith that nothing is beyond the scope of science.

A second mistake would be to underestimate the importance of the grand metaphysical story because subjects like ultimate origins are remote from the day-to-day concerns of practicing scientists. It is perfectly true that most scientists do not spend much time thinking about the ultimate metaphysical implications of the scientific enterprise, and I am sure that many are embarrassed by the hubris of the "theory of everything" school of physicists and the dogmatic materialism of the "DNA is everything" school of molecular biologists. But this is beside the point. Most people, whether they are scientists or not, go about day-to-day life without thinking about metaphysics, but their thinking is nonetheless influenced by metaphysical assumptions. In fact, metaphysical assumptions are most powerful when they are unconscious and do not come to the surface because everyone in the relevant community takes them for granted.

Relatively few scientists explicitly advocate the grand

metaphysical story, but any scientist who explicitly challenged it would quickly earn a reputation as an eccentric. Francis Crick's hypothesis may be astonishing to the general public, but when I spoke on a panel at a huge neuroscience convention, I found that his basic premise was unreflectively taken for granted by almost everyone I met. Some neuroscientists are modest about what they expect to achieve in the foreseeable future, although an expansive optimism was more in evidence, but I saw no willingness to challenge in principle the premise that mind, as part of life, *is* matter and *only* matter. After all, what else could it be?

The grand metaphysical story is therefore important, and it deserves to be taken seriously as a metaphysical story. The question is whether we have good reason to believe that the story is true, or at least more probably true than the rival metaphysical story that we were created by a supernatural being called God who cares about what we do and gives ultimate meaning to our lives.

When this question comes to the surface, which it rarely does except when scientists are debating creationists, the answer that scientific naturalists usually give is that science's past successes justify a continuing confidence in the metaphysical vision that inspired those successes. This is potentially a good argument (I will call it the argument from success), but not all kinds of success are equally relevant.

The crudest version of the argument from success relies on the technological achievements of science, like airplanes, nuclear bombs, antibiotics, and computers. From this it is only a short step to the contemptuous argument that persons who dispute the grand metaphysical story of science ought to try to travel by flying carpet or to cure their children's illnesses by faith healing. One of anthropologist Donald Johanson's popular books made the point in a memorable non sequitur: "You can't accept one part of science because it brings you good things like electricity and penicillin and throw away another part because it brings you some things you don't like about the origin of life." That reasoning overlooks the important consideration that all statements made in the name of science are not equally reliable. We believe in the efficacy of electricity and penicillin on the basis of experimental verification; many of us disbelieve claims that scientists know how life originated because we know how inadequate the experimental evidence is to justify those claims. To insist that claims be tested and not just promoted as fact because they are made by persons labeled "scientists" is simply to insist that the scientific method be followed and not just counterfeited.

In any case, the technological achievements of science have very little relevance to the vast theoretical scenarios of cosmology and evolutionary biology. Very likely the public is impressed by what theoretical physicists say about the origin of the universe mainly because an earlier generation of physicists invented the bomb that destroyed Hiroshima. But that is a fact about the public, not a fact about the origin of the universe.

The success that really matters for confirming the grand metaphysical story is the degree of success that has been achieved by historical science itself. If the scientists have actually confirmed many of the most important elements of the story, so that only a few gaps remain to be filled, then there are solid grounds (short of absolute proof) for believing that the story itself is fundamentally correct. For example, many scientists freely concede that the origin of life is still an unsolved mystery and that the materialistic theory of mind is only a hypothesis. On the other hand, even these scientists are usually confident that the neo-Darwinian theory of evolution is fundamentally correct and that science therefore knows in principle how today's complex plants and animals, including humans, evolved from the simplest life-forms by natural selection. Granted this Darwinian premise, and granted the validity of the astronomers' model of stellar evolution and the formation of the solar system and the earth, scientists have good reason to be confident that naturalistic solutions to the origin of life and consciousness problems exist to be found.

It is conceivable that God intervened twice in cosmic history, once to create the first life and subsequently to insert human consciousness into a hominid, leaving everything in between to naturalistic evolution, but scientists who think they have succeeded so brilliantly in solving most of the puzzle understandably are not impressed by a "God of the gaps" who seems likely to be replaced before long by another successful naturalistic theory.

Suppose, however, that we were to learn that the accepted theory of biological evolution is fundamentally untrue. Suppose that the Darwinian mechanism of mutation and selection cannot really create complex organs and organisms from simple beginnings, and that the problem of biological complexity has not been solved after all. If an error of that magnitude had to be confessed, the entire part of the grand metaphysical story that deals with the history and nature of life would be called into question. The confidence scientists feel that they can eventually provide a materialistic explanation for the origin of life and for consciousness would have no basis once its essential Darwinian foundation was removed. Why devote prodigious effort to speculating about how a primitive form of RNA might be produced in a chemical soup if you have no idea how such a molecule could evolve into a cell? Why assume that mind is only matter if you have no idea of how the brain could have evolved? Instead of a generally satisfactory picture of the history of life with a few gaps, science would confront a vast mystery that would become increasingly stark with the gathering of more biological data. When we imagine the consequences that would follow from a discrediting of the Darwinian theory, it is easy to understand why scientists defend the theory so fiercely.

Modernism rests on the grand metaphysical story of science, and the degree to which the story has been successfully told rests largely on the Darwinian theory of evolution. For scientific naturalists the story and the theory are virtually sacrosanct, but a theistic realist can afford to take a critical look at both. And so we will.

In general, my impression is that we are profoundly in agreement.

NOTES

The full title of Stephen Hawking's best-seller is *A Brief History of Time: From the Big Bang to Black Holes* (New York: Bantam, 1988). Quotations used in this chapter are from pages 175, 49, 136, and 12–13. The quotation from Carl Sagan's introduction appears on page x.

Hawking dedicated the book to "Jane," the heroic wife who nursed him through his disabling illness and raised his children. Jane Hawking does not appear in the excellent made-for-television movie of the book, because by that time their marriage had ended. London *Sunday Times* reporter Bryan Appleyard interviewed the Hawkings in 1988, and anyone reading Jane's candid comments about Stephen's philosophy could have guessed that trouble was brewing.

"There's one aspect of [Stephen's] thought that I find increasingly upsetting and difficult to live with," Jane told Appleyard. "It's the feeling that, because everything is reduced to a rational, mathematical formula, that must be the truth. There doesn't seem to be room in the minds of people who are working on these things for other sorts of inspiration."

When Jane tried to ask Stephen whether his theory of a cosmos with no beginning and no need for God was advanced merely as a mathematical model, or as the truth itself, she could get no reply other than the famous Hawking grin. Appleyard commented, "For Mrs. Hawking, a devout Anglican, it seems like an agnostic door slamming in her face." At another point in the interview, commenting on Stephen's growing fame, Jane remarked that her role was no longer to care for a sick man but "simply to tell him he is not God." (From Bryan Appleyard, "A Master of the Universe: Will Stephen Hawking Live to Find the Secret?" *Sunday Times* [London], July 3, 1988.)

I mention Jane Hawking's opinion not because it involves marital difficulties, which can happen to anyone, but because her criticism is extremely perceptive. Hawking, like other scientific metaphysicians discussed in this book, has an evident need to wrap reality up into a package that can be fully understood by the kind of logic that his science can employ. Whatever cannot be understood that way is pushed out of reality—however important it may be to the business of living. Who is better qualified to criticize a man's simplistic rationalism and hubris than his wife?

John Polkinghorne, a physicist, Anglican priest, and president of Queens' College of Cambridge University, provides valuable reflections on science and theology in his 1993 Gifford Lectures, published as *The Faith of a Physicist* (Princeton, N.J.: Princeton University Press, 1994). Polkinghorne characterizes Hawking's "no beginning point" hypothesis as "scientifically interesting but theologically insignificant," because "the idea of creation has no special stake in a datable start to the universe.... God is not a God of the edges, with a vested interest in boundaries. Creation is not something he did fifteen billion years ago, but it is something he is doing now" (73).

This fails to take account of the role the mathematical elimination of a beginning point plays in shoring up the confidence of the scientific naturalists, a confidence that was clearly disturbed by the prospect of having to acknowledge a scientifically ascertainable moment of creation. The "no beginning point" concept is *not* scientifically interesting: it is a mathematical construct that has no empirical basis, makes no predictions, and generates no research agenda. Its sole purpose is to support the metaphysical principle that nature is self-contained and effectively eternal.

It is true that theologians can adjust to even the most complete naturalistic system by responding, as Polkinghorne does, that God is "the sustainer of the self-contained space-time egg and the ordainer of its quantum laws," but naturalists perceive such reasoning as defensive and indicating merely that theism is unfalsifiable. Scientific naturalists do not think it necessary or possible to prove the nonexistence of an undetectable sustainer and lawgiver. All they aspire to do is to complete the naturalistic agenda as far as observable reality is concerned and leave everything beyond that to the subjective imagination. The Bible's opening words—"in the beginning God created the heavens and the earth"—state a proposition very different from a modernist theologian's concept of an undetectable sustainer of a self-contained space-time egg that has no beginning. When naturalists have forced theologians to retreat from the former to the latter, they have accomplished their purpose.

David Lindley's *The End of Physics: The Myth of a Unified Theory* (New York: BasicBooks, 1993) is an excellent corrective to the expansive scientism of the "theory of everything" school of particle physicists and cosmologists. Lindley predicts that the "end of physics" will come not because physicists will have discovered the coveted grand unified physical theory but because, long before that point, their theories will have gone so far beyond experimental testing that physics will be a branch of aesthetics more than of experimental science. The theories will seem "beautiful" in the eyes of the physicists, but they will not be testable.

Paul Davies's *The Mind of God: The Scientific Basis for a Rational World* (New York: Simon & Schuster, 1992) unabashedly presents physics as metaphysics and seems to end in pantheistic mysticism. The quotation in the text is from page 40.

Francis Crick explains his materialist starting point on page 3 of *The Astonishing Hypothesis: The Scientific Search for the Soul* (New York: Scribner's, 1994). His dismissal of religion and philosophy occurs on page 258. Crick wears his materialism on his sleeve and is refreshingly candid in expressing his contempt for the alternatives.

The quotation attributed to Arthur Kornberg is from his lecture "The Two Cultures: Chemistry and Biology," delivered at the annual meeting of the American Association for the Advancement of Science in Chicago in 1987. It was published in *Biochemistry* 26 (1987): 6888–91.

The quotation attributed to Donald Johanson is from Maitland A. Edey and Donald C. Johanson, *Blueprints: Solving the Mystery of Evolution* (Boston: Little, Brown, 1989), p. 2.

TWELVE

CRITIQUE OF PHILLIP E. JOHNSON

ROBERT T. PENNOCK

Phillip E. Johnson is professor of law at the University of California, Berkeley, and prior to his entry into the creationism debate he was best known as the author of a popular textbook on criminal law. In his book *Darwin on Trial* (1991), Johnson renewed the creationist attack against evolution, and he is rightly credited with giving the movement a new lease on life. Christian creationist groups have been quick to recognize Johnson as an important asset, and they sponsor forums for him to present his arguments against evolution. The Ad Hoc Origins Committee, a group of professors and academic scientists from universities including Princeton and the University of Texas who describe themselves as Christian Theists, claims that Johnson has given a "penetrating and fundamental critique of modern Darwinism" and distributes free videotapes of one of his speeches.[1] Creationists have become increasingly well funded and well organized in the last two decades, but until now they have lacked an articulate spokesman with a high-profile institutional affiliation. Johnson fills this role and provides the movement with the measure of credibility it has longed for. Of course, Johnson is a lawyer who boasts that he is "entirely unprejudiced because [he has] no formal training in science past high school.[2] His credibility is thus not on the scientific side—indeed, William Provine and other biologists have called his descriptions of evolutionary theory a "crude caricature"[3]—but he knows how to draw upon his strengths and makes a classic courtroom move of shifting the locus of argument in a way that seeks to undermine the expert testimony of his scientist adversaries. His key argument is broadly philosophical, but Johnson also uses his considerable rhetorical skills to try to turn the tables on scientists by portraying them as naïvely doctrinaire and intolerant, while portraying creationists as rational and fair-minded skeptics. To meet Johnson's challenge, we must not only show how his argument fails on logical grounds, but also cut through his rhetoric.

One of Johnson's titles—"Evolution as Dogma: The Establishment of Naturalism"—neatly captures both his argumentative and rhetorical strategies. Unlike the creation scientists, who try to put creationism on a par with the theory of evolution by claiming that creationism is scientific, Johnson tries to put them on a par by alleging that evolution is ideological. Darwinian evolution, he claims, "is based not upon any incontrovertible empirical evidence, but upon a highly controversial philosophical presupposition."[4] That presupposition is naturalism. Johnson argues that naturalistic evolution is not scientific but rather is a dogmatic belief system held in place by the authority of a scientific priesthood, and that without the naturalist assumption evolutionary theory would be rejected in favor of creationism. The charge that science is a "secular religion" is not new, but Johnson is the first to locate a basis for the charge in specific philosophical assumptions made by science, and to try to exploit this as a point of weakness in evolutionary theory to the advantage of creationism. Johnson's attack contains a kernel of truth—it is true that science makes use of a naturalistic philosophy—but Johnson has misunderstood naturalism's role in science in general and its implications in this instance. To show this, I will begin with a review of Johnson's main argument and discussion of its key concepts. In the course of discussion I will also highlight Johnson's prejudicial and misleading rhetoric, which serves to polarize the debate and undermine the possibility of peaceful coexistence between science and religion.

JOHNSON AGAINST THE "DOGMA OF NATURALISM"

Johnson's Argument

Johnson offers variations of the usual creationist arguments that try to poke holes in the broad fabric of scientific evidence for evolution, but we will focus upon his novel and strongest challenge, which is the whole-cloth charge that evolution is metaphysical dogma. We can summarize Johnson's main argument in the following three-step form: Evolution is a naturalistic theory that denies by fiat any supernatural intervention. The scientific evidence for evolution is weak, but the philosophical assumption of naturalism dogmatically disallows consideration of the creationist's alternative explanation of the biological world. Therefore, if divine interventions were not ruled out of court, creationism would win over evolution.

This is not laid out formally as a deductive argument, but one recognizes at once a version of the familiar "dual model" tactic; the argument is presented as though evolution and creationism are the only alternatives, so if evolution gets knocked out, creationism wins by default. Creation scientists, requesting "balanced treatment" of the issue in the public schools, used a very crude form of this type of argument structure, with Darwinian evolution on the one side and a thinly disguised biblical literalism on the other; let the children judge the evidence and decide for themselves which one is right, they asked in the name of fairness. (Of course, they did not plan to mention Mayan or Hindu or Asanti creation stories as alternatives.) Johnson is more sophisticated. He, too, wants to get his conclusion by means of a negative argument against evolution,[5] but he tries harder to set up the dichotomy to logically exclude other alternatives by attempting to define the key terms of the debate—"Darwinism" and "creationism"—so that they are mutually exclusive and jointly exhaustive.

Johnson takes pains to distinguish his brand of creationism from the specific scripture-based commitments of creation science[6] and to define creationism broadly. Here is the way he puts the definition in *Darwin on Trial*:

> "Creationism" means belief in creation in a . . . general sense. Persons who believe that the earth is billions of years old, and that simple forms of life evolved gradually to become more complex forms including humans, are "creationists" if they believe that a supernatural Creator not only initiated this process but in some meaningful sense *controls* it in furtherance of a purpose.[7]

Elsewhere he reiterates this with a slightly different emphasis:

> The essential point of creation has nothing to do with the timing or the mechanisms the Creator chose to employ, but with the element of design or purpose. In the broadest sense, a "creationist" is simply a person who believes that the world (and especially mankind) was *designed,* and exists for a *purpose.*[8]

A significant feature of Johnson's definitions is that they put no explicit restrictions on the manner of creation so long as God is involved in a significant way; guided evolution, special creation, or any other mode of divine creation seems allowed. The definitions make no reference to the Bible, making it appear that Johnson countenances as creationist the cosmogonies of any other religious or cultural tradition. In *Evolution as a Dogma,* Johnson is even more general:

> [A] "creationist" is . . . any person who believes that God creates.[9]

Such apparent open-mindedness makes the defender of evolution look narrow-minded in contrast with the tolerant creationist. It also serves to enlarge Johnson's constituency, for most people will identify themselves as creationist in the minimal sense of commitment to the idea that God creates.[10] Additionally, the broad definition helps bolster Johnson's claim that evolution is necessarily at odds with religion, for he contrasts this mild-mannered creationism with a view of evolutionary theory that makes the latter essentially atheistic.

Johnson defines "evolution" very narrowly. He does not deny that evolution by natural selection occurs if all one means by that is that "limited changes occur in populations due to differences in survival rates."[11] Even creation science allows microevolution, he claims—God created "kinds" but thereafter individuals can diversify within the limits of the kind.[12] On the other hand, the important thesis of evolutionary theory, he says, is the further one about macroevolution: that evolutionary processes also explain "how moths, trees, and scientific observers came to exist in the first place."[13] Most of *Darwin on Trial* attacks this claim, but Johnson narrows his sights still further to set up his general argument. Since it is possible that God did not create creatures suddenly, but used instead a gradual evolutionary process, even macroevolution does not contradict creationism unless it is "explicitly or tacitly defined as *fully naturalistic evolution*—meaning evolution that is not directed by any purposeful intelligence."[14] This is the form of evolution that Johnson sets up as his target. Here is his positive definition:

> By "Darwinism" I mean fully naturalistic evolution, involving chance mechanisms guided by natural selection.[15]

Johnson's main argument hangs on his conception of the role of naturalism in this scheme, which we will examine shortly, but his central point is that in naturalistic evolution God's intervention is excluded.

Taking these two definitions together, we see how the argument is supposed to work. Creationism holds that God plays a role in Creation (however it occurs) and Darwinism denies the same. Though closer inspection makes it clear that Johnson's definitions do not establish the logical dichotomy he needs, on the surface it looks as though he has set up the major premise of a valid dilemma that will then allow creationists to rely solely upon negative argumentation. This is Johnson's first innovation. His second is that his characterization of the terms of the debate allows evolution to be attacked not only on scientific grounds, but also on philosophical grounds. He spends seven chapters in *Darwin on Trial* on the first task, trying to cast doubt upon the wide range of empirical evidence for evolutionary theory so that he can claim that creationism is a better theory which would be accepted if not for the "powerful" and "doctrinaire" naturalistic assumption that rules it out by definition.[16] As we shall see, however, the philosophical charge does the real work.

Johnson is well aware that scientists have not and will not now find creationist criticisms of the evidence for evolution to be persuasive,[17] but this matters little for he is playing to the jury. When the scientific expert witness rebuts his negative appraisals of the evidence for evolution, Johnson will argue that biologists are "[unable] to make any sense out of creationist criticisms of their presuppositions" because of their "philosophical naïveté"[18] and their "blind commitment to naturalism."[19] The evidence cited for evolution, Johnson claims,

> looks quite different to people who accept the possibility of a creator outside the natural order. To such people, the peppered-moth observations and similar evidence seem absurdly inadequate to prove that natural selection can make a wing, an eye, or a brain. From their more skeptical perspective, the consistent pattern in the fossil record of sudden appearance followed by stasis

> tends to prove that there is something wrong with Darwinism, not that there is something wrong with the fossil record. The absence of proof "when measured on an absolute scale" is unimportant to a thoroughgoing naturalist, who feels that science is doing well enough if it has a plausible explanation that maintains the naturalistic worldview. The same absence of proof is highly significant to any person who thinks it possible that there are more things in heaven and earth than are dreamt of in naturalistic philosophy.[20]

Again we see how Johnson's rhetoric tries to make the creationist appear to be the rational "skeptic" who merely accepts the "possibility" of a Creator, and the biologist the "blind" and "naïve" ideologue who dogmatically rejects that possibility and thereby misjudges the evidence. Thus, he concludes, it is not the evidence but the ideology that supports evolution. Here is the conclusion the reader is supposed to draw:

> Victory in the creation-evolution dispute therefore belongs to the party with the cultural authority to establish the ground rules that govern the discourse. If creation is admitted as a serious possibility, Darwinism cannot win, and if it is excluded a priori Darwinism cannot lose.[21]

The claim that evolution is held up solely by "metaphysical assumptions"[22] and "speculative philosophy"[23] allows Johnson to ignore the weakness of his negative scientific arguments. In his public lectures, Johnson follows the same pattern, usually taking a few token swipes at the empirical evidence for evolution and then moving quickly to his philosophical indictment of its naturalistic metaphysics.

Although this philosophical criticism of naturalism has to carry the weight of his conclusion, Johnson fails to provide any philosophical analysis of the concept that he charges scientists have uncritically accepted. Neither does he support his thesis that the concept is inherently dogmatic, or provide evidence that scientists do subscribe to it in the way he claims. Let us now briefly review the history of naturalism, and then evaluate Johnson's characterization and his application of the concept to the biological case.

Varieties of Naturalism

The generic meaning of "naturalism" is a philosophical view based upon study of the natural world, with an implicit contrast to the supernatural world, but this leaves room for a wide range of specific variations. Since the time of the ancient Greeks, naturalism has often been associated with various forms of secularism, especially epicureanism and materialism, but it has also been used as a label for religious views such as pantheism, as well as the theological doctrine that we learn religious truth not by revelation but by the study of natural processes. In the centuries leading up to the twentieth century, concomitant with the rise of the natural sciences, the term became associated more directly with the methods and fruits of the scientific study of nature. One spin-off at the turn of the century was the naturalist movement in literature, epitomized by Zola but continuing in a form through Steinbeck, which featured "scientific" portrayals of human characters playing out predetermined roles as amoral creatures governed by natural law. Another extreme expression was Auguste Comte's philosophy of *positivism*, the scientific stage of philosophical development which society purportedly reached after progressing beyond theological and metaphysical conceptions of the world. Positivism concerned itself only with regularities of observable phenomena, so naturalism at that time became associated with phenomenalism. This version of naturalism was carried forward into the philosophy of science in the early twentieth century by the influential logical positivists, who restricted knowledge to propositions with a determinable truth-value—if a proposition was not verifiable then it was taken to be meaningless. The so-called verifiability criterion of meaning turned out to be unworkable, and its collapse was one of several reasons for the demise of the logical positivist view in the middle of the century. Since then, in philosophy at least, the naturalist view of the world has become coincident with the scientific view of the world, whatever that may turn out to be. Many people continue to think of the scientific world view as being exclusively materialist and deterministic, but if science discovers forces and fields and indeterministic causal processes, then these too are to be accepted as part of the naturalistic worldview.[24] The key point is that naturalism is not necessarily tied to specific ontological claims (about what sorts of being do or don't exist); its base commitment is to a method of inquiry.

Of course one could choose to take some set of basic ontological categories from science at a particular time and then claim that only these things exist. The seventeenth-century mechanistic materialists, who held that the world consists of nothing but material particles in motion, did just this, and there are any number of other ways that one could decide to fix base ontological commitments. This type of view is known as *metaphysical* or *ontological naturalism.* The ontological naturalist makes substantive claims about what exists in nature and then adds a closure clause stating "and that is all there is." A thorough historical review of positive formulations of ontological naturalism could fill an article in itself, but amidst this variety many do agree on a common negative claim: because God standardly is assumed to be supernatural, the typical ontological naturalist denies God's existence. It is possible, however, for an ontological naturalist to allow God in the picture, provided God's attributes are appropriately constrained to conform to the regimen of the given natural ontology. Hobbes and Spinoza were ontological naturalists who thought they found room for God (indeed, for a Judeo-Christian God) in this way. Some traditional theists, however, were not willing to countenance their naturalized conceptions of the deity; Hobbes was branded an atheist and Spinoza a pantheist. The problem of trying to naturalize theology is that traditionalists want God to be able to control nature from outside nature; they take God to be supernatural by definition. Probably the main reason for the strong secularist strand among the varieties of naturalism is that many naturalists also have tended to take for granted this traditional conception of God and have found it difficult to square with their other ontological commitments.

Ontological naturalism should be distinguished from the more common contemporary view, which is known as *methodological naturalism.* The methodological naturalist does not make a commitment directly to a picture of what exists in the world, but rather to a set of methods as a reliable way to find out about the world—typically the methods of the natural sciences, and perhaps extensions that are continuous with them and indirectly to what those methods discover. An important feature of science is that its conclusions are defeasible on the basis of new evidence, so whatever tentative substantive claims the methodological naturalist makes are always open to revision or abandonment on the basis of new, countervailing evidence. Because the base commitment of a methodological naturalist is to a mode of investigation that is good for finding out about the empirical world, even the specific methods themselves are open to change and improvement; science might adopt promising new methods and refine existing ones if doing so would provide better evidential warrant. Understanding the nature of scientific evidence is critical for answering Johnson's charge, but let us postpone examination of that concept and how it relates to the question of God's existence and creativity until we have seen the details of Johnson's philosophical claims that naturalism is assumed dogmatically and that its ideology alone supports evolutionary theory.

Although it is the linchpin of his argument, Johnson provides only a cursory discussion of the concept of naturalism. Taken individually, his few statements do pick out versions of naturalism, but taken together they suggest a biased and misleading picture. In *Darwin on Trial,* Johnson defines naturalism as follows:

> Naturalism assumes the entire realm of nature to be a closed system of material causes and effects, which cannot be influenced by anything from "outside." Naturalism does not explicitly deny the mere existence of God, but it does deny that a supernatural being could in any way influence natural events, such as evolution, or communicate with natural creatures like ourselves.[25]

This is a good definition of a common form of ontological naturalism; the "causal closure of the physical" is another way this idea is expressed. The acknowledgment that naturalism does not "explicitly" deny the "mere existence" of God, however, is significant, for it is another indication that Johnson is not as tolerant and ecumenical as his definition of creationism might initially lead one to believe. The clear implication here is that, because naturalism rejects continuing divine intervention, it does *implicitly* deny God's existence, but this conclusion follows only if one has a particular conception of divine power. We see this view expressed again as Johnson immediately follows the above definition by introducing a specific form of naturalism that he calls "scientific naturalism."

> Scientific naturalism makes the same point by starting with the assumption that science, which studies only the natural, is our only reliable path to knowledge. A God who can never do anything that makes a difference, and of whom we can have no reliable knowledge, is of no importance to us.[26]

Note that in such statements Johnson is dismissing views such as Deism that do allow God to influence natural events, to make a difference and conceivably even to communicate with us, by setting up the world in the appropriate way at Creation but thereafter not intervening in the natural order. He is also rejecting views that hold that God is concerned with our spiritual rather than our material being and thus intervenes only at a spiritual level. He is also ignoring religious views that do not posit a personal God, but conceive of God as a universal life force or a mystical unity. Also unimportant, apparently, are views that say we can have "no reliable knowledge" of God; this restriction leaves out even many Judeo-Christian thinkers who hold that the nature of God is unknowable to the human mind. These spiritual views Johnson excludes are prevalent worldwide, so we should not be misled by his attempt to portray his form of creationism as generically tolerant. Such views, however, *are* compatible with varieties of both ontological and methodological naturalism and belie Johnson's attempts to conflate naturalism and atheism.

Returning to the definitions, one may think at first that "scientific naturalism" is Johnson's term for methodological naturalism, but in light of his other comments we see that he mixes in elements of ontological naturalism. He says, for example, that in the present context he considers scientific naturalism to be equivalent to evolutionary naturalism, scientific materialism, and scientism:

> All these terms imply that scientific investigation is either the exclusive path to knowledge or at least by far the most reliable path, and that only natural or material phenomena are real. In other words, what science can't study is effectively unreal.[27]

By ignoring distinctions among such positions[28] Johnson again is able to associate evolution with (godless) materialism and to portray naturalism as monolithically dogmatic. "Scientism," for example, is a term of derision coined by hermeneutic critics of science to label those who wanted to apply the methods of the natural sciences "inappropriately" to the human sciences, for which they thought the literary model of hermeneutic *interpretation* should reign as the proper method. Their target was specifically the followers of the logical positivists, but, as was noted, the exclusionary positivist view that only the scientifically verifiable was meaningful has not held currency for several decades. Contemporary methodological naturalists would not recognize themselves in this description, yet it is just this sort of view that Johnson insistently portrays as the essence of scientific naturalism.[29]

When he applies naturalism to evolution Johnson says that one gets:

> a theory of naturalistic evolution, which . . . absolutely rules out any miraculous or supernatural intervention at any point. Everything is conclusively presumed to have happened through purely material mechanisms that are in principle accessible to scientific investigation, whether they have yet been discovered or not.[30]

Here it is clear that Johnson is describing a form of ontological naturalism—besides the reference to mechanistic materialism, the terms "absolutely" and "conclusively" emphasize supposed dogmatic commitment to the substantive ontological claims. Johnson claims that evolutionary biologists assume this sort of positivistic philosophy, but certainly evolutionary biology as a science does not have to do so, and it is hard to believe even that any scientist who has kept abreast of developments in philosophy of science would affirm this form of ontological naturalism.

Indeed, it seems clear that the two biologists that Johnson most often decries—George Gaylord Simpson and Stephen Jay Gould—do not endorse such a view, but are instead methodological naturalists. Simpson discussed naturalism as part of his review of the principle of uniformitarianism in geology and biology and is explicit that the scientific postulate of naturalism is "a necessity of method" and that the rejection of appeal to preternatural factors must be made on "heuristic grounds."[31] When one looks for Gould's view on the matter, one finds in his discussion of uniformitarianism that he used precisely the distinction reviewed above to disambiguate "substantive uniformitarianism" (a descriptive hypothesis holding that the history of life was uniform) from "methodological uniformitarianism."[32] Gould uses the latter term to label the assumption in geology that natural laws are invariable—a position that implies absence of supernatural intervention. The name Gould gives this presupposition tells us just how he views it; he recommends that the special term be dropped because it follows from the fact that geology is a science. Clearly, both Simpson and Gould understand that science does not affirm naturalism as a substantive ontological claim but rather as a methodological assumption.

METHODOLOGICAL NATURALISM AND EVIDENCE

We have seen how Johnson misleadingly inserts terminology with connotations of dogmatism into the definition of naturalism. He regularly refers to naturalism using such terms as "extravagant extrapolation, arbitrary assumptions, and metaphysical speculation,"[33] but such name-calling is no argument. Johnson provides no analysis to show that science assumes the naturalistic principle dogmatically; he simply asserts this. We have now seen that naturalism is not properly put forward as an ontological claim about what conclusively does or does not exist, but rather as a methodological rule that states a valid way for investigation to proceed, so clearly it is not dogmatic in the sense Johnson claimed. But is the methodological rule itself dogmatic? To say that a belief or principle is dogmatic is to say that it is opinion put forward as true or valid on the grounds of authority rather than reason. Does science put forward the methodological principle not to appeal to supernatural powers or divine agency simply on authority? Is it just an extravagant, arbitrary, speculative assumption? Certainly not. There is a simple and sound rationale for the principle based upon the requirements of scientific evidence.

Empirical testing relies fundamentally upon the lawful regularities of nature which science has been able to discover and sometimes codify in natural laws. For example, telescopic observations implicitly depend upon the laws governing optical phenomena. If we could not rely upon these laws—if, for example, even when under the same conditions, telescopes occasionally magnified properly and at other occasions produced various distortions dependent, say, upon the whims of some supernatural entity—we could not trust telescopic observations as evidence. The same problem would apply to any type of observational data. Lawful regularity is at the very heart of the naturalistic worldview and to say that some power is supernatural is, by definition, to say that it can violate natural laws.[34] So, when Johnson argues that science should allow in supernatural powers and intelligences he is in effect saying that it should allow beings that are above the law (a rather strange position for a lawyer to take). But without the constraint of lawful regularity, inductive evidential inference cannot get off the ground. Controlled, repeatable experimentation, for example, which Johnson explicitly endorses in his video "Darwinism on Trial" (1992), would not be possible without the methodological assumption that supernatural entities do not intervene to negate lawful natural regularities.

Of course, science is based upon a philosophical system, but not one that is extravagant speculation. Science operates by empirical principles of observational testing; hypotheses must be confirmed or disconfirmed by reference to empirical data. One supports a hypothesis by showing that consequences obtain which would follow if what is hypothesized were to be so in fact. As we have seen, Darwin spent most of *Origin of Species* applying this procedure, demonstrating how a wide variety of biological phenomena could have been produced (and thus are explained) by the simple causal processes of the theory. Supernatural theories, on the other hand, can give no guidance about what follows or does not follow from their supernatural components. For instance, nothing definite can be said about the processes that would connect a given effect with the will of the supernatural agent—God might simply say the word and zap anything into or out of existence. Furthermore, in any situation, any pattern (or lack of pattern) of data is compatible with the general hypothesis of the existence of a supernatural agent unconstrained by natural law. Because of this feature, supernatural hypotheses remain immune from disconfirmation.[35] Johnson's form of creationism is particularly guilty on this count. Creation science does include supernatural views at its core that are not testable, and it was rightly dismissed as not being scientific because of these in the Arkansas court case, but it at least was candid about a few specific nonsupernatural claims that are open to disconfirmation (and indeed that have been disconfirmed), such as that the earth is less than 10,000 years old and that many geological and paleontological features were caused by a universal flood (the Noachian Deluge). Johnson, however, does not provide any creationist claim beyond his generic one that "God creates for some purpose," and, as a purely supernatural hypothesis, this is not open to empirical test. Science assumes methodological naturalism because to do otherwise would be to abandon its empirical evidential touchstone.

Finally, allowing appeal to supernatural powers in science would make the scientist's task just too easy, because one would always be able to call upon the gods for quick theoretical assistance. Johnson wants us to accept the claim that "God creates for some purpose" as an explanation of the biological world, but there would be no reason to stop there. Once such supernatural explanations are permitted, they could be used in chemistry and physics as easily as creationists have used them in biology and geology. Indeed, all empirical investigation beyond the purely descriptive could cease, for scientists would have a ready-made answer for everything. Obviously, science must reject this kind of one-size-fits-all explanation. By disqualifying such short-cuts, the naturalist principle also has the virtue of spurring deeper investigation. If one were to find some phenomenon that appeared inexplicable according to some current theory one might be tempted to attribute it to the direct intervention of God, but methodological naturalism prods one to look further for a natural explanation. Clearly, it is not just because such persistence has proven successful in the past that science encourages this attitude.

Johnson claims that "if the possibility of an 'outside' intervention is allowed in nature at any point . . . the whole naturalistic world-view quickly unravels."[36] He intends by this only that atheistic Darwinism will lose in a head-on comparison with theistic creationism once the "ideological" restrictions are removed but, as we have seen, the consequences would be far more serious. Johnson wants to make an exception to the law in this one area, but it would infect the entire enterprise. Methodological naturalism is not a dogmatic ideology that simply is tacked on to the principles of scientific method; it is essential for the basic standards of empirical evidence.

Creationism's Evidence

With his attack upon naturalism, Johnson is arguing that science abandon a sound methodological principle and reintroduce miraculous "explanations." We have seen that science has good reasons for retaining this principle—without it, standard inductive evidential inferences would be undermined—but we have also admitted that rules of scientific inquiry are themselves open to change or modification if a better method of evidential warrant is found. Does Johnson have something better in mind? It seems that he does, for he regularly claims that the supernatural theory of creationism is a better theory than Darwinism, and he constantly complains that "creationists are disqualified from making a positive case, because science by definition is based upon naturalism."[37] Such statements lead one to expect that Johnson will supply what he says the scientific priesthood has suppressed, but one will look in vain for this positive evidence. Amidst all the negative arguments one finds only two small hints of what type of positive evidence the creationists have to offer—revelation and the design argument.

The first occurs only as a passing remark following an (inadvertently self-undermining) acknowledgment that empiricism is a "sound methodological premise."[38] Johnson writes:

> Science is committed by definition to . . . find[ing] truth by observation, experiment, and calculation rather than by studying sacred books or achieving mystical states of mind. It may well be, however, that there are certain questions . . . that cannot be answered by the methods available to our science. These may include not only broad philosophical issues such as whether the universe has a purpose, but also questions we have become accustomed to think of as empirical, such as how life first began or how complex biological systems were put together.[39]

The sly implication here is that the "sacred books" and "mystical states of mind" could indeed be appropriate ways to answer empirical as well as teleological questions. Is this Johnson's new source of positive evidence for creationism? I asked Johnson just this question following one of his public lectures and he replied that he was not defending this position. However, neither did he deny that such appeal to scriptural authority or mystical experience would count as positive empirical evidence. Johnson seems to be pleading the Fifth on this important issue. He cannot reject these methods without alienating his constituency, for the biblical account, perhaps supplemented by religious experiences, is the prime motivation for Christian creationists. On the other hand, he cannot endorse the "method" of supernatural revelation without abdicating his claim of expertise as a lawyer, for anyone would be laughed out of court who argued that one could help establish an empirical fact (say, that the defendant set off the explosion) by reference to the authority of psychic or spiritual testimony.

The second hint of a source of positive evidence for creationism is in the following statement:

> [F]rom a creationist point of view, the very fact that the universe is on the whole orderly, in a manner comprehensible to our intellect, is evidence that we and it were fashioned by a common intelligence.[40]

This is the only instance where Johnson makes an explicit commitment to any type of positive evidence the creationist can provide, but what we have here is nothing but a version of the old *argument from design*—the world appears to exhibit a designed arrangement so we should infer the existence of a designer—which relies on, at best, only a weak analogy from the human case to the divine. This is not the place to review the vast literature on the design argument, so I will confine myself to a few remarks on Johnson's specific formulation of it.

What sort of order do we find in the universe, and what can it tell us? Examples of design which we attribute to a human designer include such things as a house, a formal well-manicured garden, or Paley's famous pocketwatch, but I would argue that we draw the inference in these cases precisely because the kind of order we see in them is so *unlike* what we typically find in nature; their simple geometrical forms and periodicities are strikingly different from the complexities and irregularities that the surface structure of the world presents. And when we do discover an underlying order in nature that is "comprehensible to our intellect," it is order to which we have been able to give natural, scientific explanations. Such order thus does not provide good reason to infer a supernatural creator. It is rather those features of the world that, for the time being at least, are *incomprehensible* to our intellect that are more likely to lead us to think of a higher power. At one time this might have been the awful power of a thunderstorm, leading us to suppose an angry god. Or we might have been struck by the wondrous beauty of a rainbow after the storm and interpreted it as a sign that God was appeased. We typically appeal to supernatural agency to explain that which we cannot explain otherwise. But when these phenomena are eventually accounted for in terms of natural electrical and optical properties, they lose their persuasiveness as indications of the literal presence of God and at most retain only an emotional or symbolic force.

Similarly, the version of the design argument that appeals to the adaptedness of organisms was persuasive only when this adaptation was mysterious and the idea of purposeful creation seemed the only possible explanation. But Darwin showed how simple natural processes could explain such adaptations. Johnson's "God creates for a purpose" view can say nothing about the supernatural processes by which the Creation was accomplished or what divine ends it serves; such an "explanation" starts and stops with the will of God. To pick just one of any number of common examples, Darwin's theory also accounts for those organisms that are not properly adapted to their environment—random variation produces both fit and unfit individuals, and natural selection is more likely to eliminate those that cannot compete as well in the given environment—but why would God create the world in such a way that the vast majority of individual organisms die because they are maladapted? The design argument has always been criticized on this sort of point even before Darwin; if God designed the world, why did he do such a poor job of it? Isn't the evident waste and sloppiness actually an argument *against* the existence of God? Such questions show the flip side of the design argument and highlight its weakness, for if it is applicable at all it is applicable for both the theist and the atheist. The creationist answer to such impious questions is that God must have had his reasons. Period.

There are two ways a creationist might become more specific about the will and methods of God. The first would be to appeal to revelation through mystical experience or scripture, but we have already seen that this does not in itself count at all as empirical evidence. The other is to revert to an earlier form of naturalism we mentioned—natural theology—and to try to judge the nature of God by looking at the book of nature. Johnson indulges explicitly in this approach just once, concluding that the elaborate-tailed peacock and peahen are "just the kind of creatures a whimsical Creator might favor, but that an 'uncaring mechanical process' like natural selection would never permit to develop."[41] If we are to take Johnson seriously, such "creationist explanations" in terms of divine "whimsy" postulated on the basis of peacock tails, are better accounts than those given by evolutionary theory and are thereby supposed to favor creationism.

If this is the "positive case" that purportedly has been suppressed, it is no wonder that creationists rely exclusively upon negative argumentation and why Johnson labors so mightily to legitimize it. The creationists' insistence upon viewing the issue as a simple either/or choice mistakenly leads them to think that their negative arguments against Darwinian evolutionary mechanisms directly prove creationism. However, negative argument will not suffice to establish the creationists' desired conclusion even using Johnson's nonstandard definitions of "Darwinism" and "creationism" that were quoted earlier. Johnson's definition of Darwinism mentioned only the classical evolutionary mechanism of natural selection operating upon tiny chance variations, and omitted other possible processes. So even if negative argument were to rule out this sort of mechanism (something that Johnson certainly has not done), one could not thereby accept creationism since there are other alternatives that do not rely upon divine intervention. His definitions of creationism are similarly problematic. For example, as mentioned above, one definition seems to rule out the sort of Deist who believes that God created the world and set it going in the way he wanted, but then no longer intervenes. Another definition rules out supernatural but nontheistic views that would stand in contrast to naturalistic evolution. Johnson would have us believe that the logical situation is the following:

Creationism (C_1) or Darwinism (D_1)

as though these were the only candidates. If this were the case, and a negative argument disproved the latter, then the former would follow as a deductive logical conclusion, but the logical situation is rather:

$$C_1 \text{ or } C_2 \text{ or} \ldots \text{or } C_n \text{ or } D_1 \text{ or } D_2 \text{ or} \ldots \text{or } D_n \text{ or } X_n,$$

and in this case, even if D_1 were rejected, a variety of options remains besides Johnson's preferred C_1, so purely negative argument is not sufficient to establish it. Furthermore, these positions are not necessarily mutually exclusive. Direct conflict occurs when evolutionary theory is confronted with *specific* creationist stories (like the creation scientists' literal six-day instantaneous creation), but Johnson claims not to defend any such view.

"Darwinism," as Johnson defines it at least, is not a single thesis but rather the conjunction of at least two different theses: (D_m) the specific evolutionary mechanism of the modern synthesis . . . and (D_a) the atheistic denial of divine intervention. But all of Johnson's and other creationists' negative arguments are directed at undermining (D_m), so even if they were successful they would leave the possibility of (D_a) untouched. The existence or nonexistence of some particular evolutionary process is independent of the question of whether or not there is a creative deity. Johnson himself must grant this for he claims to allow the possibility that "he might have created things instantaneously in a single week or through gradual evolution over billions of years."[42] Thus, negative arguments against evolutionary processes are irrelevant to the key question of divine creative power when stated in this general way. Indeed, if Johnson cared only about his broad, ecumenical sort of creationism as he claims, then he should have no reason at all to argue against (D_1) and could confine himself to arguing against (D_2). Only someone who has a specific conflicting creationist scenario in mind, such as the one-week instantaneous Creation story, need worry about the evolutionary mechanism. In any case, negative argumentation is not going to establish either the general or the specific creationist thesis. As prosecuting attorney for the new creationists, Johnson needs to provide positive evidence for his and his clients' preferred conclusion, but, as we have seen, he has none to offer.

METHODOLOGICAL NATURALISM VERSUS THEISTIC SCIENCE

Unsurprisingly, Johnson and other IDCs [Intelligent Design Creationists] are as unhappy with methodological naturalism as they are with metaphysical naturalism. When Johnson published a response to my argument of the previous two sections he argued that the former is just a slick sales ploy to sucker the unwary into buying the latter. He wrote:

> That distinction implies that Darwinists do not claim to make ontologically true statements about the history of life, but only statements about what inferences can be drawn from a naturalistic starting point. If the Darwinists were really as modest as that, there would be little to argue about. For example, I agree with Richard Dawkins that the "blind watchmaker" mechanism is the most plausible naturalistic hypothesis for how complex organisms might have come into existence. We disagree only over whether the theory is true.[43]

He goes on to say:

> We who know how this game of bait-and-switch is played just look for the "switch" that turns innocent "methodological" naturalism into the real thing. In Pennock's version, the switch is the argument that naturalism and rationality are virtually identical—because he thinks that attributing the design of organisms to an intelligence would imply that all events occur at the whim of capricious gods, so that there would be no regularities for scientists to observe. No doubt this caricature explains why no science was done in the century of Newton.[44]

But Johnson's statements about truth and rationality are misleading. Johnson wants truth about reality—indeed, he wants absolute truth—but he neglects the more basic issue that truth claims must be justified by some method. He says it is irrational to act in conflict with true reality and asserts that God is truly real, but he fails to provide a method by which we can justify claims about God. Instead, he challenges scientists to say that they know, rather than just dogmatically assume, that God did not create us. But we have seen that biologists do not assert by fiat that God played no role in the development of life-forms; they simply proceed, as all scientists must, to search for purely natural mechanisms. When they find evidence for a natural explanation (and Johnson admits that the Darwinian mechanism is not only possible but somewhat plausible), they can legitimately say that they have discovered something true about the natural world. To be sure, this is an approximate and tentative scientific truth, not an ontological (metaphysical) truth, in the sense that it cannot rule out the possibility that a supernatural Creator is involved in the process. (On the other hand, it does rule out one version of the teleological argument: that God is necessary to explain this development.)

Consider for comparison the geneticist who, applying methodological naturalism, searches for a natural explanation for hypertrichosis. People with hypertrichosis grow hair all over their faces and upper bodies, and were once thought of as werewolves. Finding evidence for the X-linked gene and an evolutionary explanation of the trait, the geneticist might reassure a patient that his disorder is "the result of a purposeless and natural process that did not have him in mind," the phrase of G. G. Simpson that creationists find so offensive. Surely we may accept that statement as true, even though as a merely naturalistic scientific truth, it does not rule out the possibility of an intelligent supernatural cause—a "curse of the werewolf," say—so it cannot be said to be absolutely true in the ontological (metaphysical) sense. Similarly, the creationists' supernatural story might be a metaphysical truth—God might have created the world 6,000 years ago but made it look older as mature-earth "appearance of age" creationists hold—but it is not a scientific truth.

In the most important section of *Reason in the Balance* Johnson proposes "theistic realism" and tries to support the creationist hypothesis from the viewpoint of its "theory of knowledge."[45] He begins by citing John 1:1–13 as "the essential, bedrock position of Christian theism about creation,"[46] and he goes on to make his central argument that it is "obvious" to all who have not had their reason clouded by the "drug" of naturalism that living beings are the products of intelligent creation: "Because in our universal experience unintelligent material processes do not create life."[47] Weigh that reason in the balance! Genetic engineering might one day allow humans to create life, but so far we do not have a single case of intelligent creation of life; rather, our universal experience to date is that *only* unintelligent material processes do so. Here the important point is that the idea of a theistic science is set out in opposition to methodological naturalism.

Intelligent-design creationists unite in this attack and on their insistence in the viability of theistic science. Johnson continues to write as though methodological naturalism is essentially synonymous with metaphysical naturalism, but others acknowledge that methodological naturalism is a distinct view and attack it directly. Notre Dame philosopher of religion Alvin Plantinga, for example, joins the IDCs in opposing evolution and in rejecting methodological naturalism. As he puts it, "A Christian academic and scientific community ought to pursue science in its own way, *starting from* and taking for granted what we know as Christians."[48] In arguing against methodological naturalism and in championing such "Augustinian science," Plantinga admits that his suggestion "suffers from the considerable disadvantage of being at present both unpopular and heretical" but argues that it has the "considerable advantage of being correct."[49] Plantinga suggests that a theistic scientist could reason as follows:

> God has created the world, and of course has created everything in it directly or indirectly. After a great deal of study, we cannot see how he created some phenomenon *P* (life, for example) indirectly; thus probably he has created it directly.[50]

But we have already considered this form of argument. When God is brought in to explain what we find unexplainable we have no more than a God of the gaps, and this form of reasoning is little different from the negative argumentation of the dual model. Plantinga takes pains to distance his epistemology from the God-of-the-gaps view, but he can do so only by explicitly giving up the idea that God is meant to be an explanatory hypothesis and by appealing to revelation.

> First, the thought that there is such a person as God is not, according to Christian theism, a hypothesis postulated to *explain* something or other, nor is the main reason for believing that there is a such a person as God the fact that there are phenomena that elude the best efforts of current science. Rather, our knowledge of God comes by way of *general* revelation, which involves something like Aquinas's general knowledge of God or Calvin's *sensus divinitatis* and also, and more importantly, by way of God's *special* revelation, in the Scriptures and through the church.[51]

Johnson's "theological science"[52] might emphasize different Scripture and Plantinga's "Augustinian science" might have a more sophisticated theology, but with regard to the creationism debate neither is too different from the standard "creation science" of Henry Morris's Bible-based young-earth creationism in the sense that all these begin with some assumption of what Christians supposedly "know" and can take for granted in their science. But the battles we witnessed within the Tower over what may be presumed as "True Christianity" with regard to the "theological facts" should give us sufficient reason to doubt whether revelation could possibly supply the purported unified basis for such a science. Broaden one's view to observe battlegrounds of theological disputes among competing religious traditions and even angelic scientists would fear to tread there.

We also hear echoes of Morris in Johnson's comment about Newtonian science, in which he follows a trend among creationists to pick up on studies in the history of science that show how some natural philosophers of the scientific revolution supported their notion of natural law by reference to a conception of God as an orderly law-giver. Creationists are now eager to credit Christianity with the origin of modern science (or at least of the "True Science" that supports their views) and to anachronistically call theistic scientists like Newton fellow "creationists." However, they want Newton as one of their own because of his scientific successes, not because of his less enduring theological work on "ancient wisdom," interpretation of the Book of Revelations, and numerology regarding "the number of the beast." God might have underwritten the active principles that govern the world described in the *Principia* and the *Opticks,* but he did not interrupt any of the equations or regularities therein. Johnson and other creationists who want to dismiss methodological naturalism would do well to consult Newton's own rules of reasoning, especially his first that says not to admit unnecessary causes when explaining phenomena, and his fourth that says to regard the conclusions of inductive methods as "accurately or very nearly true"[53] and to eschew contrary hypotheses until new evidence requires them. Are such rules metaphysical dogma? Commentators note that Newton's *theology* sometimes led him to regard these dogmatically, but that in his scientific passages he took them as "a matter of method merely, to be used tentatively as a principle of further inquiry."[54]

In Newton's day, many Christians thought atomism was tantamount to atheism in much the same way that Johnson and other creationists now say that evolution is, and Newton engaged in spirited exegesis to combat this view. He traced the atomic idea back through pagans such as Lucretius, Epicurus, Democritus, Thales, Pythagoras, and finally to Mochus, whom he identified with Moses to make the link to Christianity. Today Christian theists rightly find such contrivances unnecessary, for theism is not threatened by atomic theory. Nor do Christians still feel troubled by the heliocentric and geokinetic theories of the solar system, though an earlier generation went through a great turmoil over that conflict with the Bible. Such theories have been confirmed under the methodological assumptions of naturalistic science and we may properly call them true and factual in as strong a sense as empirical justification allows; they are not metaphysical but simple mundane truths. Evolutionary theory is of a kind and, again after some turmoil, most religious groups have come to accept it as a scientific truth about reality that is fully compatible with their faith and with a mature understanding of Scripture. Johnson's, Plantinga's and other IDCs' attempts to turn back the clock do a disservice to both religion and science.

I certainly agree that believers can be good scientists; what I cannot agree with is that they must be believers of the creationist variety who want not just absolute truth, but their unjustified, anti-scientific version of it. Scientific naturalism is no dogma. It is a sound principle of method.

NOTES

1. Andrew Bocarsly et al., Ad Hoc Origins Committee's Letter to Colleagues, 1993.

2. Gil Silberman, "Phil Johnson's Little Hobby," *Boalt Hall Cross-Examiner* 6, no. 2 (1993): 9.

3. William B. Provine, "Response to Phillip Johnson," in *Evolution as Dogma: The Establishment of Naturalism,* ed. Phillip E. Johnson (Dallas: Haughton, 1990), p. 20.

4. Phillip E. Johnson, ed., *Evolution as Dogma: The Establishment of Naturalism* (Dallas: Haughton, 1990), p. 1.

5. One of Johnson's main complaints against science, which he illustrates by reference to the National Academy of Science's "friend of the court" brief in the Arkansas trial, is that it does not allow creationism to use merely negative arguments. Phillip E. Johnson, *Darwin on Trial,* (Washington, D.C.: Regnery Gateway, 1991), p. 8.

6. As we have seen, politically organized creationists are mostly Christian literalists who believe, for example, in biblical inerrancy, in special Creation by God of all biological kinds, and in a historical global Flood. The Bible-Science Association, the Creation-Science Fellowship Inc., and the Creation Research Society, among others, all require that one sign a statement of belief in such principles to qualify for membership.

7. Johnson, *Darwin on Trial,* p. 4.

8. Ibid., p. 113

9. Johnson, *Evolution as Dogma,* p. 13.

10. Barbara J. Culliton, "The Dismal State of Scientific Literacy," *Science* 243 (February 1989): 600, reports that only half of Americans accepted the statement that "human beings as we know them today developed from earlier species of animals." Of these many will agree that this evolutionary development was guided by God.

11. Johnson, *Evolution as Dogma,* p. 2.

12. Johnson notes the idea that the different races of human beings are all descendants of the original ancestral pair—Adam and Eve (*Evolution as Dogma,* p. 68). He calls this an example of allowable microevolution, but clearly the Creationist notion of diversification "within the limits" of created "kinds" is not the same as microevolution on the biological model. Furthermore, it is doubtful that creation scientists, given their theological beliefs, could consistently accept that races differentiated on the basis of natural selection of random genetic variation.

13. Johnson, *Evolution as Dogma,* p. 2.

14. Johnson, *Darwin on Trial,* p. 4.

15. Ibid.

16. Johnson, *Evolution as Dogma,* p. 6.

17. The standard creationist arguments that Johnson brings out have previously been addressed by scientists and philosophers such as Philip Kitcher, *Abusing Science: The Case against Creationism* (Cambridge: MIT Press, 1982); Michael Ruse, *Darwinism Defended* (Reading, Mass.: Addison-Welsley, 1982); Arthur N. Strahler, *Science and Earth History: The Creation/Evolution Controversy* (Amherst, N.Y.: Prometheus Books, 1987); and Tim M. Berra, *Evolution and the Myth of Creationism* (Stanford: Stanford University Press, 1990).

18. Johnson, *Evolution as Dogma,* p. 13.

19. Ibid., p. 17.

20. Ibid., p. 8.

21. Ibid., p. 8.

22. Ibid., p. 13

23. Ibid., p. 15.

24. To recognize the enrichment of the scientific ontology beyond the classical form of materialism, it is now more common to speak of "physicalism," where the reference is to the ontology of current physics, or sometimes "physicalistic materialism." Admittedly, some scientists and philosophers continue to use simple "materialism," but it is understood in the broader sense of the fabric of the universe, which includes space-

time and electromagnetic fields, and so on, rather than in the old sense of mere *matter.* Johnson, however, explicitly links naturalism and the old mechanistic materialism throughout his work, with the rhetorical effect of conflating naturalistic evolution with atheistic materialism.

25. Johnson, *Darwin on Trial,* pp. 114–15.

26. Ibid., p. 115.

27. Ibid., p. 114.

28. Johnson provides no bibliographic references for any of these terms so we cannot evaluate the specifics of the definitions he may have in mind.

29. Another example is the passage: "The Soviet Cosmonaut who announced upon landing that he had been to the heavens and had not seen God was expressing crudely the basic philosophical premise that underlies Darwinism. Because we cannot examine God in our telescopes or under our microscopes, God is unreal. It is meaningless to say that some entity exists if in principle we can never have knowledge of that entity (Johnson, *Evolution as Dogma,* p. 14.). Not only does Johnson once again link naturalism to atheism here by way of communism, but he explicitly characterizes it in positivist terms, with the reference to the verifiability criterion of meaning.

30. Johnson, *Evolution as Dogma,* p. 3.

31. George Gaylord Simpson, "Uniformitarianism: An Inquiry into Principle, Theory, and Method in Geohistory and Biohistory," in *Essays in Evolution and Genetics in Honor of Theodosius Dobzhansky,* ed. M. K. Kecht and W. C. Steere (New York: Appleton-Century-Crofts, 1970), p. 61.

32. Stephen J. Gould, "Is Uniforitarianism Necessary?" *American Journal of Science* 265 (March 1965): 226.

33. Johnson, *Evolution as Dogma,* p. 15.

34. This is not to say, however, that things we now think of as supernatural necessarily are so. It could turn out, for example, that ghosts exist but that unlike our fictional view of them, they are subject to natural law. In such a case we would have learned something new about the natural world (which may require revising current theories), and would not have truly found anything supernatural.

35. I should note a possible red herring on this point. Creationists often erroneously claim that evolutionary theory itself is not disconfirmable, and so they charge that it is in the same boat as their view. To his credit, Johnson does not make this error. He understands that the theory is disconfirmable, and all of his negative argumentation purports to show

that in fact it has already been disconfirmed. Johnson's positive claim is that biologists' philosophical prejudices prevent them from recognizing the disconfirming evidence.

36. Johnson, *Evolution as Dogma,* p. 7.

37. Ibid., p. 9.

38. Ibid., p. 14.

39. Ibid.

40. Ibid., p. 13

41. Johnson, *Darwin on Trial,* p. 31.

42. Ibid., p. 113.

43. Philip E. Johnson, "Response to Pennock," *Philosophy of Biology* 11 (1996): 561.

44. Ibid.

45. Philip E. Johnson, *Reason in the Balance: The Case against Naturalism in Science, Law, and Education* (Downer's Grove, Ill.: InterVarsity Press, 1995), p. 107.

46. Ibid.

47. Ibid., p. 108.

48. Alvin Plantinga, "Methodological Naturalism?" *Perspectives on Science and Christian Faith* 49, no. 3 (1997): 143.

49. Ibid., pp. 143–44.

50. Ibid., p. 154.

51. Ibid., p. 149.

52. Johnson, *Reason in the Balance,* p. 105.

53. Isaac Newton, *Mathematical Principles of Natural Philosophy and His System of the World,* trans. Andrew Motte, rev. Florian Cajori (1729; reprint, Berkeley: University of California Press, 1962), p. 399.

54. E. A. Burtt, *The Methaphysical Foundations of Modern Science* (New York: Harcourt, Brace, & Co., 1925), p. 215.

THIRTEEN

SCIENCE OR BEAUTY?

MARTIN GARDNER

My previous column ("Literary Science Blunders," January–February 1995) was about the gulf that too often divides the culture of science from the culture of the liberal arts. Nowhere is this chasm more noticeable than in the lines of certain poets who believe that a knowledge of science somehow destroys one's awareness of the wonders and beauty of nature.

Over the decades I've collected some examples:

Reprinted from "Science vs. Beauty," *Skeptical Inquirer* (March/April 1995). Used by permission of the *Skeptical Inquirer*—www.psicop.com.

The moon shines down with borrowed light,
So savants say—I do not doubt it.
Suffice its silver trance my sight,
That's all I want to know about it.
A fig for science. . . .

—Robert Service

The goose that laid the golden egg
Died looking up its crotch
To find out how its sphincter worked.
Would you lay well? Don't watch.

—X. J. Kennedy, "Ars Poetica"

I'd rather learn from one bird how to sing
Than teach ten thousand stars how not to dance.

—e e cummings

While you and i have lips and voice which
are for kissing and to sing with
who cares if some one eyed son of a bitch
invents an instrument to measure Spring with?

—e e cummings

. . . Do not all charms fly
At the mere touch of cold philosophy?
There was an awful rainbow once in heaven:
We know her woof, her texture; she is given
In the dull catalogue of common things.
Philosophy will clip an Angel's wings,
Conquer all mysteries by rule and line,
Empty the haunted air and gnomed mine—
Unweave a rainbow. . . .

—John Keats, "Lamia"

"Arcturus" is his other name—
I'd rather call him "Star."
It's very mean of Science
To go and interfere!

I pull a flower from the woods—
A monster with a glass
Computes the stamens in a breath
And has her in a "class"!

—Emily Dickinson

A Color stands abroad
On Solitary Fields
That Science cannot overtake
But Human Nature feels.

It waits upon the Lawn,
It shows the furthest Tree
Upon the furthest Slope you know
It almost speaks to you.

Then as Horizons step
Or Noons report away
Without the Formula of sound
It passes and we stay—

A quality of loss
Affecting our Content
As Trade had suddenly encroached
Upon a Sacrament.

—Emily Dickinson

Sweet is the lore which Nature brings;
Our meddling intellect
Mis-shapes the beauteous forms of things:—
We murder to dissect.

Enough of Science and of Art;
Close up those barren leaves;
Come forth, and bring with you a heart
That watches and receives.

—Wordsworth, "The Tables Turned"

Today we breathe a commonplace,
Polemic, scientific air;

We strip illusion of her veil;
We vivisect the nightingale
To probe the secret of his note
The Muse in alien ways remote
    Goes wandering.

—Thomas Bailey Aldrich

Science! true daughter of Old Time thou art!
    Who alterest all things with thy peering eyes.
Why preyest thou thus upon the poet's heart,
    Vulture, whose wings are dull realities?

—Edgar Allan Poe

There's machinery in the butterfly,
There's a mainspring to the bee.
There's hydraulics to a daisy
And contraptions to a tree.

If we could see the birdie
That makes the chirping sound
With psycho-analytic eyes,
With X-ray, scientific eyes,
We could see the wheels go round.

And I hope all men
Who think like this
Will soon lie underground.

—Vachel Lindsay

Similar sentiments have been expressed in prose. Here are a few: Coleridge: "The real antithesis of poetry is not prose but science." Billy Rose: "I wish the engineers would keep their slide rules out of the bits of fairyland left in this bollixed up world." Nietzsche: "They [scientists] have cold, withered eyes before which all birds are unplumed."

There is something to be said for such sentiment, though not much. It is possible for scientists to become so wrapped up in their work that they lose all sense of nature's beauty and mystery. "When you understand all about the sun and all about the atmosphere and all about the rotation of the earth," wrote Alfred North Whitehead, a philosopher who stood astride the two cultures, "you may still miss the radiance of the sunset."

G. K. Chesterton made the same point in his amusing story "The Unthinkable Theory of Professor Green," in *Tales of the Long Bow*. Green is an astronomer who forgot about the world around him until one day when he fell in love with a farmer's daughter. He announces a lecture on his discovery of a new planet. The auditorium is packed with colleagues while he describes one of the planet's strange creatures. Slowly it dawns on Green's listeners that he is describing a cow.

I suppose that scientists like Professor Green, before he discovered the earth, exist, but if so, I have yet to encounter one. On the contrary, almost all scientists believe that as their knowledge increases, their sense of wonder also grows. The scientist sees a flower, said physicist John Tyndall, "with a wonder superadded."

Professor Green's unthinkable theory reminds me of stanza xcii from the first canto of Byron's *Don Juan*:

> He thought about himself, and the whole earth,
>   Of man the wonderful, and of the stars,
> And how the deuce they ever could have birth;
>   And then he thought of earth-quakes, and of wars,
> How many miles the moon might have in girth,
>   Of air-balloons, and of the many bars
> To perfect knowledge of the boundless skies;—
> And then he thought of Donna Julia's eyes.

The Laputans, in *Gulliver's Travels*, describe a woman's beauty by "rhombs, circles, parallelograms, ellipses, and other geometrical terms." Arthur S. Eddington, writing on "Science and Mysticism" in *The Nature of the Physical World*, quotes from a page on winds and waves in a textbook on hydrodynamics. He then compares this with the aesthetic experience of watching actual sea waves "dancing in the sunshine."

That knowledge of science adds to one's appreciation of the mystery and splendor of the cosmos has nowhere been more vigorously expressed than by the late physicist Richard Feynman, in Christopher Syke's *No Ordinary Genius* (W. W. Norton, 1994). He described an artist friend who would hold up a flower and say: "I, as an artist, can see how beautiful a flower is. But you, as a scientist, take it all apart and it becomes dull."

"I think he's kind of nutty," says Feynman, and he adds:

> First of all, the beauty he sees is available to other people—and to me too. Although I might not be quite as refined aesthetically as he is, I can appreciate the beauty of a flower.
>
> At the same time, I see much more about the flower than he sees. I could imagine the cells in there, the complicated actions inside, which also have a beauty. I mean, it's not just beauty at this dimension of one centimeter: there is also beauty at a smaller dimension—the inner structure. The fact that the colors in the flower are evolved in order to attract insects to pollinate it is interesting—it means that the insects can see the color. It adds a question: does this aesthetic sense also exist in the lower forms? Why is it aesthetic? All kinds of interesting questions which a science knowledge only adds to the excitement and mystery and the awe of a flower. It only adds. I don't understand how it subtracts.
>
> Does it make any less of a beautiful smell of violets to know that it's molecules? To find out, for example, that the smell of violets is very similar to the chemical that's used by a certain butterfly (I don't know whether it's true, like my father's stories!), a butterfly that lets out this chemical to attract all its mates? It turns out that this chemical is exactly the smell of violets with a small change of a few molecules. The different kinds of smells and the different kinds of chemicals, the great variety of chemicals and colors and dyes and so on in the plants and everywhere else, are all very closely related, with very small changes, and the efficiency of life is not always to make a new thing, but to modify only slightly something that's already there, and make its function entirely different, so that the smell of violets is related to the smell of earth. . . . These are all additional facts, additional discoveries. It doesn't take away that it can't answer questions of what, ultimately, does the smell of violets really feel like when you smell it. That's only if you expected science to give the answers to

> every possible question. But the idea that science takes away is something I don't understand.
>
> It's true that technology can have an effect on art that might be a kind of subtraction. For example, in the early days painting was to make pictures when pictures were unavailable, that was one reason: it was very useful to give people pictures to look at, to help them think about God, or the Annunciation, or whatever. When photography came as a result of technology, which itself was the result of scientific knowledge, then that made pictures very much more available. The care and effort needed to make something by hand which looked exactly like nature and which was once such a delight to see now became mundane in a way (although of course there's a new art—the art of taking good pictures). So yes, technology can have an effect on art, but the idea that it takes away mystery or awe or wonder in nature is wrong. It's quite the opposite. It's much more wonderful to know what something's really like than to sit there and just simply, in ignorance, say, "Oooh, isn't it wonderful!"

A famous poem by Walt Whitman tells how he listened to a "learn'd astronomer" lecture about the heavens until he (Walt) became "unaccountably tired and sick." He walks out of the lecture room into the "mystical moist night air" so he can look up "in perfect silence at the stars."

Here is how Feynman, in his *Lectures on Physics*, reacted to the notion that astronomical knowledge dulls one's sense of awe toward the cosmos:

> "The stars are made of the same atoms as the earth." I usually pick one small topic like this to give a lecture on. Poets say science takes away from the beauty of the stars—mere gobs of gas atoms. Nothing is "mere." I too can see the stars on a desert night, and feel them. But do I see less or more? The vastness of the heavens stretches my imagination—stuck on this carousel my little eye can catch one-million-year-old light. A vast pattern—of which I am a part—perhaps my stuff was belched from some forgotten star, as one is belching there. Or see them with the greater eye of Palomar, rushing all apart from some common starting point

> when they were perhaps all together. What is the pattern, or the meaning, or the *why?* It does not do harm to the mystery to know a little about it. For far more marvelous is the truth than any artists of the past imagined! Why do the poets of the present not speak of it? What men are poets who can speak of Jupiter if he were like a man, but if he is an immense spinning sphere of methane and ammonia must be silent?

Isaac Asimov, writing on "Science and Beauty" in *The Roving Mind*, quotes Whitman's poem. "The trouble is that Whitman is talking through his hat," says Asimov. Of course the night sky is beautiful, but is there not a deeper, added beauty provided by astronomy? Asimov continues with lyrical paragraphs about the "weird and unearthly beauty" of our sister planets, as recently disclosed by space probes, about the awesome wonders of the stars, of the billions of galaxies each containing billions of suns, of clusters of galaxies, and superclusters fleeing from each other as the universe expands from its incredible origin in the explosion of a tiny point some 15 billion years ago.

> And all of this vision—far beyond the scale of human imaginings—was made possible by the works of hundreds of learn'd astronomers. All of it; *all* of it was discovered after the death of Whitman in 1892, and most of it in the past twenty-five years, so that the poor poet never knew what a stultified and limited beauty he observed when he look'd up in perfect silence at the stars.
>
> Nor can we know or imagine now the limitless beauty yet to be revealed in the future—by science.

STUDY QUESTIONS

Johnson seems to think that the physicists' quest for a theory of everything (TOE) is an effort to exclude the concept of God as creator. Suppose that scientists do succeed in formulating and confirming a TOE. Would this exclude God? In his notes at the end of his essay, Johnson admits that a TOE would still leave room for God as the sustainer of the cosmos and as the reason why we have one ultimate set of laws as opposed to another. Why does he find that answer objectionable? Does his reply show that it is not merely the existence of God he wishes to protect from naturalistic science but a particular biblical or religious conception?

Johnson argues that the effort of neuroscientists to provide a naturalistic theory of the human mind will reduce mind to a mere "sideshow" in nature. He thinks that this project is self-defeating because its success would show that unthinking processes determine all our reasoning, including that of neuroscientists. Is this argument sound? Why cannot rational thought be a physical process?

Pennock argues that science must pursue a naturalistic methodology that considers only natural hypotheses. Is this true? Could a supernaturalistic hypothesis be scientifically evaluated? As noted in the introduction, in the final chapter of *Origin of Species,* Darwin compares natural selection and special creation head-to-head against the evidence. He concludes that special creation does make some predictions—which are, in fact, false. Instead of excluding supernatural hypotheses on any a priori grounds, even methodological ones, why not just proceed to test them vis-à-vis the evidence?

REFERENCES AND FURTHER READING

A BIBLIOGRAPHICAL ESSAY

This essay gives the bibliographical details of the works from which the readings were drawn and other works that pertain to the issues. It is not in any way a comprehensive bibliography. It merely lists some of the works I have found most useful in learning about these issues. I comment on some texts; the reader may or may not find these comments helpful.

The founding document of the social constructivist movement is *Laboratory Life: The Construction of Scientific Facts* by Bruno Latour and Steve Woolgar, first published in 1979 by Sage Publications and reprinted in 1986, with a new introduction and index, by Princeton University Press. It is a pretty tough read, both because of the technical subject matter and, to be honest, because the authors do not always seem to strive for maximum clarity. Actually, the constructivist thesis they present could be interpreted in a number of ways. A sympathetic, thorough, and readable study of constructivism as it relates to the history of science is Jan Golinski's *Making Natural Knowledge: Constructivism and the History of Science* published by Cambridge University Press in 1998. Golinski argues that constructivism should not be seen as making extreme claims that deny the reality of the physical world or that science is severed from that reality. Golinski says that constructivism is merely the attempt to focus on the human aspects of science. However, some of Latour and Woolgar's statements in *Laboratory Life* seem to belie such an innocuous construal. Woolgar's 1988 book *Science: The Very Idea* (London: Tavistock) is in fact quite hostile to all scientific claims to represent reality and even to the claim that there is some objective reality to be represented.

Latour has modified his position several times since *Laboratory Life*. His more recent books include *Science in Action* (Cambridge: Harvard University Press, 1987) and *We Have Never Been Modern* (Cambridge: Harvard University Press, 1993). In some of these later works Latour, who now evinces great indignation if he is characterized as a "constructivist," urges that we make "one more turn" after the social turn. That extra turn is, in the eyes of some readers, a turn back toward some degree of realism (See Sergio Sismondo *Science without Myth* [Albany: State University of New York Press, 1996]). Others doubt whether Latour's "creeping realism" (Sismondo's phrase) is sincere or consistent.

Shapin and Schaffer's *Leviathan and the Air-Pump* (Princeton: Princeton University Press, 1985) is another of the most important constructivist studies. Shapin and Schaffer are good historians who have a solid grasp of the social circumstances of Restoration Britain. Also, the book is well written and interesting. The authors are wholly committed to a constructivist interpretation, which unsympathetic readers will inevitably see as ideological blinders.

Robert Klee's *Introduction to the Philosophy of Science: Cutting Nature at its Seams* (Oxford: Oxford University Press, 1997) is an excellent introduction to the philosophy of science. It is particularly valuable in that it pays much attention to social constructivism, feminism, and other recent views on science that have followed on the publication of Thomas Kuhn's *The Structure of Scientific Revolutions,* 2d ed. (Chicago: University of Chicago Press). Anyone wanting to understand the "social turn" in the study of science needs to read Kuhn.

Constructivism has been a very successful movement. Constructivists completely dominate the sociology of science and to a great extent the history of science also. A dissenting sociologist is Stephen Cole, whose book *Making Science: Between Nature and Society* was published by Harvard University Press in 1992. Though Cole rejects the old-fashioned positivist view of science, he also provides a thorough critique of constructivism. He has a more sharply worded rebuttal in the volume edited by Gross, Levitt, and Lewis (see below) titled "Voodoo Sociology: Recent Developments in the Sociology of Science."

Critiques of constructivism by philosophers include two by James Robert Brown: *The Rational and the Social* (London: Routledge, 1989), and *Smoke and Mirrors: How Science Reflects Reality* (London: Routledge, 1994). Brown, though himself politically left-wing, deplores the adoption of constructivist and relativist positions by left/liberal academics. Alan Chalmers's *Science and Its Fabrication* (Minneapolis: University of Minnesota Press, 1990) also criticizes constructivism by focusing on constructivists' own case studies and disputing their interpretations. A book that uses case studies from the history of dinosaur paleontology to criticize constructivism and postmodernism is my own (Parsons's) *Drawing Out Leviathan: Dinosaurs and the Science Wars* (Bloomington: Indiana University Press, 2001).

Feminist philosophers have written a great deal about science in the last twenty years. Two important early works are the anthology edited by S. Harding and M. Hintikka, *Discovering Reality: Feminist Perspectives on Epistemology, Metaphysics, Methodology, and Philosophy of Science,* (Dordrecht: Reidel, 1983) and E. F. Keller's, *Reflections on Gender and Science* (New Haven: Yale University Press, 1985). Sandra Harding has been a very prolific contributor to the feminist philosophy of science. Her work *Whose Science? Whose Knowledge? Thinking from Women's Lives* (Ithaca, N.Y.: Cornell University Press, 1991) provided the selection included in this book. One of the most cited works in feminist philosophy of science is Helen Longino's *Science as Social Knowledge: Values and Objectivity in Scientific Inquiry* (Princeton, N.J.: Princeton University Press, 1990).

Of the various critiques of the feminist philosophy of science, one of the best known is one included in this volume—Cassandra Pinnick's "Feminist Epistemology: Implications for the Philosophy of Science" from the journal *Philosophy of Science* 61 (1994): 646–57. The volume edited by Gross, Levitt, and Lewis contains several responses to feminist philosophy. Two particularly interesting ones are Janet Radcliffe Richards's "Why Feminist Epistemology Isn't" (pp. 385–412) and Noretta Koertge's "Feminist Epistemology: Stalking an Un-Dead horse" (pp. 413–19). A book-length critique of feminist epistemology and philosophy of science is Ellen R. Klein's *Feminism Under Fire* (Amherst, N.Y.: Prometheus Books, 1996) from which the selection by Klein in this book was taken. One very interesting part of Klein's book discusses the difficulties critics of feminist philosophy have had in getting included in conferences on the topic or published in anthologies dealing with it.

Postmodernist writings are particularly difficult to understand, and the movement as a whole seems to revel in the arcane. A clear, simple introduction to postmodernism is G. Ward's *Postmodernism* (Chicago: NTC/Contemporary Publishing, 1997), one of the "Teach Yourself" books. Though lambasted by Cartmill, Donna Haraway's *Primate Visions: Gender, Race, and Nature in the World of Modern Science* (New York: Routledge, 1989) is actually one of the most clearly written works of this genre. The Internet has some interesting material on Haraway, especially an interview originally published in *Wired* magazine. An in-depth critique of postmodernist critical theory, deconstructionism, and relativism by someone who knows the literature very well is Christopher Norris's *Against Relativism: Philosophy of Science, Deconstruction, and Critical*

Theory (Oxford: Blackwell, 1997). Matt Cartmill's review of *Primate Visions* was first published in *International Journal of Primatology* 12, no. 1 (1991). Steven Weinberg's original article on the Sokal hoax appeared in the *New York Review of Books,* August 8, 1996, and the exchange with critics in the October 3, 1996, issue. The article and exchange are reprinted in *The Sokal Hoax: The Sham that Shook the Academy,* edited by the editors of the magazine *Lingua Franca* (Lincoln: University of Nebraska Press, 2000). This book also reproduces the original article by Alan Sokal, "Transgressing the Boundaries: Toward a Transformative Hermeneutics of Quantum Gravity," that started the whole imbroglio. *Lingua Franca* is a very interesting source for staying apprised of all the developments in academe.

There is a very considerable literature on creation vs. evolution, much of it strongly polemical one way or the other. A solid, even-handed history of creationism is Ronald L. Numbers's *The Creationists: The Evolution of Scientific Creationsim* (New York: Alfred A. Knopf, 1992). In the early 1990s, just as it seemed that the debate over creationism was dying down, Phillip E. Johnson published *Darwin on Trial* (Washington, D.C.: Regnery Gateway, 1991), which began the most recent phase of the debate—the controversy over "Intelligent Design Theory." Perhaps surprisingly, several widely respected philosophers have come out in support of this view of creationism. Alvin Plantinga, the doyen of American philosophers of religion, is perhaps the best known of these (check his Web site for details). Plantinga also joins Johnson in his attack on naturalism. He even argues that scientists who are Christians should not even respect methodological naturalism (see his essay "Methodological Naturalism?" on the Internet at http://www.arn.org/docs/odesign/od181/methnat181.htm). The selection by Johnson in this book is from *Reason in the Balance: The Case Against Naturalism in Science, Law & Education* (Downer's Grove, Ill.: InterVarsity Press, 1995). A number of Johnson's essays and interviews are also available on the Internet.

The most comprehensive reply to Intelligent Design Creationism is Robert T. Pennock's *Tower of Babel: The Evidence Against the New Creationism* (Cambridge: MIT Press, 1999). Also of great interest is Kenneth R. Miller's *Finding Darwin's God: A Scientists Search for Common Ground between God and Evolution* (New York: HarperCollins, 1999). Miller, a Brown University biologist and a devout Christian argues forcefully that creationism is wrong and that one can be a Christian and a Darwinian. See also *Can a Darwinian Be a Christian: The Relationship between Science and Religion* (Cambridge: Cambridge University Press, 2001).

Finally, the three most comprehensive works dealing with the "science wars" are Paul R. Gross and Norman Levitt's *Higher Superstition: The Academic Left and Its Quarrels with Science* (Baltimore: Johns Hopkins University Press, 1994). This no-holds-barred polemic really put the "war" into the "science wars." The critique of Shapin and Schaffer included in the present volume came from this book. The "academic left" fired back in *Science Wars,* edited by Andrew Ross (Durham, N.C.: Duke University Press, 1996). The critiques of the "academic left" continue in Gross, Levitt, and Martin Lewis, eds., *The Flight from Science and Reason* (New York: New York Academy of Sciences, 1996).

LIST OF CONTRIBUTORS

MATT CARTMILL is a professor in the Department of Biological Anthropology and Anatomy at Duke University. His publications include *Primate Origins* (1975) and *A View to a Death in the Morning* (1994).

MARTIN GARDNER has published many books since his classic *Fads and Fallacies in the Name of Science* (1952). A regular columnist for many years with *Scientific American* and the *Skeptical Inquirer,* he is one of the best-known popularizers and defenders of science.

PAUL R. GROSS is the former director of the Woods Hole Biological Laboratory and is university professor emeritus at the University of Virginia. He is the coauthor with Barbara Forrest of *Evolution and the Wedge of Intelligent Design* (2003) and the coauthor with Norman Levitt of *Higher Superstition* (1994).

DONNA HARAWAY is a professor of the history of consciousness at the University of California, Santa Cruz. Her publications include *Primate Visions* (1990) and *Simians, Cyborgs, and Women* (1991).

SANDRA HARDING is a professor of social sciences and comparative education at UCLA and the director of the UCLA Center for the Study of Women. Her publications include *The Science Question in Feminism* (1986) and *Whose Science? Whose Knowledge? Thinking from Women's Lives* (1991).

PHILLIP E. JOHNSON is the Jefferson E. Peyser Professor of Law at the University of California, Berkeley. He has lectured widely and participated in numerous debates with evolutionists. His publications include *Darwin on Trial* (1991) and *Reason in the Balance* (1995).

ROBERT KLEE is an associate professor of philosophy and chair of the Department of Philosophy and Religion at Ithaca College. He is the author of *Introduction to the Philosophy of Science: Cutting Nature at Its Seams* (1997) and editor of the anthology *Scientific Inquiry* (1999).

ELLEN R. KLEIN is an associate professor of philosophy at Flagler College. Her publications include *Feminism Under Fire* (1996).

BRUNO LATOUR is a professor at the Centre de sociologie de l'innovation Ecole des Mines in Paris and the University of California, San Diego. His publications include *Laboratory Life* (1979; with Steve Woolgar), *Science in Action* (1987), and *The Pasteurization of France* (1988).

NORMAN LEVITT is a professor of mathematics at Rutgers University. He is the author of *Prometheus Bedeviled* (1999) and coauthor, with Paul Gross, of *Higher Superstition* (1994).

ROBERT T. PENNOCK is an associate professor at the Lyman Briggs School and in the Philosophy Department of Michigan State University. He is the author of *Tower of Babel* (1999).

CASSANDRA PINNICK is a professor of philosophy in the Department of Philosophy and Religion at Western Kentucky University. She has published articles on feminist philosophy and other topics relating to the social study of science. She is coeditor with Noretta Koertge and Robert F. Almeder of *Scrutinizing Feminist Epistemology: An Examination of Gender in Science* (2003).

SIMON SCHAFFER is a reader in the Department of History and Philosophy of Science at the University of Cambridge. His publications include *Leviathan and the Air-Pump* (1985; coauthored with Steven Shapin), and *The Uses of Experiment* (coeditor; 1989).

STEVEN SHAPIN is a professor of sociology at the University of California, San Diego. His publications include *Leviathan and the Air-Pump* (1985; coauthored with Simon Schaffer) and *The Scientific Revolution* (1996).

STEVEN WEINBERG is a winner of the 1979 Nobel Prize for physics and is a professor of physics at the University of Texas at Austin. He is the author of *The First Three Minutes* (1977), *Dreams of a Final Theory* (1994), and *Facing Up* (2001).

STEVE WOOLGAR coauthored *Laboratory Life* with Bruno Latour in 1979. His other publications include *Science: The Very Idea* (1988). He has been Senior Lecturer in Sociology at Brunel University and is now director of the Virtual Society? Programme at the Said Business School of Oxford University.